Design of Visualizations for Human-Information Interaction

A Pattern-Based Framework

Synthesis Lectures on Visualization

David Ebert, *Purdue University*
Niklas Elmqvist, *University of Maryland*

Synthesis Lectures on Visualization publishes 50- to 100-page publications on topics pertaining to scientific visualization, information visualization, and visual analytics. Potential topics include, but are not limited to: scientific, information, and medical visualization; visual analytics, applications of visualization and analysis; mathematical foundations of visualization and analytics; interaction, cognition, and perception related to visualization and analytics; data integration, analysis, and visualization; new applications of visualization and analysis; knowledge discovery management and representation; systems, and evaluation; distributed and collaborative visualization and analysis.

Design of Visualizations for Human-Information Interaction: A Pattern-Based
Framework Kamran Sedig and Paul Parsons

ISBN: 978-3-031-01474-1 print
ISBN: 978-3-031-02602-7 ebook

DOI 10.1007/978-3-031-02602-7

A Publication in the Springer series
SYNTHESIS LECTURES ON VISUALIZATION #5
Series Editors: David S. Ebert, Purdue University, Niklas Elmqvist, University of Maryland

Series ISSN 2159-516X Print 2159-5178 Electronic

Design of Visualizations for Human-Information Interaction

A Pattern–Based Framework

Kamran Sedig
Western University

Paul Parsons
Purdue University

SYNTHESIS LECTURES ON VISUALIZATION #5

ABSTRACT

Interest in visualization design has increased in recent years. While there is a large body of existing work from which visualization designers can draw, much of the past research has focused on developing new tools and techniques that are aimed at specific contexts. Less focus has been placed on developing holistic frameworks, models, and theories that can guide visualization design at a general level—a level that transcends domains, data types, users, and other contextual factors. In addition, little emphasis has been placed on the thinking processes of designers, including the concepts that designers use, while they are engaged in a visualization design activity.

In this book we present a general, holistic framework that is intended to support visualization design for human-information interaction. The framework is composed of a number of conceptual elements that can aid in design thinking. The core of the framework is a pattern language—consisting of a set of 14 basic, abstract patterns—and a simple syntax for describing how the patterns are blended. We also present a design process, made up of four main stages, for creating static or interactive visualizations. The 4-stage design process places the patterns at the core of designers' thinking, and employs a number of conceptual tools that help designers think systematically about creating visualizations based on the information they intend to represent.

Although the framework can be used to design static visualizations for simple tasks, its real utility can be found when designing visualizations with interactive possibilities in mind—in other words, designing to support a human-information interactive discourse. This is especially true in contexts where interactive visualizations need to support complex tasks and activities involving large and complex information spaces. The framework is intended to be general and can thus be used to design visualizations for diverse domains, users, information spaces, and tasks in different fields such as business intelligence, health and medical informatics, digital libraries, journalism, education, scientific discovery, and others. Drawing from research in multiple disciplines, we introduce novel concepts and terms that can positively contribute to visualization design practice and education, and will hopefully stimulate further research in this area.

KEYWORDS

visualization, human-information interaction, design thinking, design intention, pattern language, design framework, design process, systems theory, information space, visual representation, interaction design, interactivity, complex activities, tasks, theory, conceptual framework, information items, information mapping, science of visualization

Contents

Figure Credits

Figure 2.1 (R): Courtesy of LHOON, http://commons.wikimedia.org/wiki/File:Sankeysteam.png.

Figure 3.1: Adapted from: Skyttner, L. (2005). General Systems Theory: Problems, Perspectives, Practice (2nd Ed.). Hackensack, NJ: © World Scientific. Used with permission.

Figure 3.3: Courtesy of © 2016 Tableau Software.

Figure 3.12: Adapted from: Cleveland, W.S. and McGill, R. (1984). Graphical perception: Theory, experimentation, and application to the development of graphical methods. Journal of the American Statistical Association, 79(387), 531–554.

Figure 3.13: Adapted from: Mackinlay, J. (1986). Automating the design of graphical presentations of relational information. *ACM Transactions on Graphics*, 5(2), 110–141.

Figure 5.1a: Courtesy of Matthew Cole/Shutterstock.com.

Figure 5.5a: With permission and adapted from: Scharein, R.G. (2015). KnotPlot [Computer software]. Vancouver, Canada: Hypnagogic Software. Available: http://knotplot.com/download/.

Figure 5.5b: Courtesy of Mayavi.

Figure 5.6a: From: Heer, J., and Bostock, M. (2010). Crowdsourcing graphical perception: using mechanical turk to assess visualization design. In Proceedings of the SIGCHI Conference on Human Factors in Computing Systems (pp. 203–212). Copyright © 2010 ACM. Used with permission.

Figure 5.6b: Courtesy of Prof. Mark Dubin, University of Colorado-Boulder.

Figure 5.6c: From: Devesa, S.S., Grauman, D.G., Blot, W.J., Pennello, G., Hoover, R.N., Fraumeni, J.F. Jr. Atlas of cancer mortality in the United States, 1950–94. Washington, DC: US Govt Print Off; 1999 [NIH Publ No. (NIH) 99–4564].

Figure 5.6d: Courtesy of iQoncept/Shutterstock.com.

Figure 5.7c: Courtesy of Martin Theus and Mondrian (http://www.theusRus.de/Mondrian).

Figure 5.8d: Courtesy of © 2015 The Regents of the University of California, through the Lawrence Berkeley National Laboratory.

Figure 5.9: Courtesy of OpenBible.info.

Figure 5.10: Copyrighted 2016. Technology Review. 121014:0216BN.

Figure 5.11a: Courtesy of Beao, https://commons.wikimedia.org/wiki/File:Chess_board_blank.svg.

Figure 5.11b: **Courtesy of Armtuk,** https://commons.wikimedia.org/wiki/File:Periodic_Table_ Armtuk3.svg.

Figure 5.11c: Courtesy of quince.infragistics.com.

Figure 5.12b: Courtesy of Max Roberts, www.tubemapcentral.com.

Figure 5.16a: Courtesy of Karen Minot.

Figure 5.16b: Courtesy of New Scientist.

Figure 5.17: Courtesy of Jan Jongert, Nels Nelson, Anna Brambilla, Cyclifiers.org, Rotterdam, 2010.

Figure 5.18: Adapted from: Benton, M. J. (1998). The quality of the fossil record of vertebrates. In Donovan, S. K. and Paul, C. R. C. (Eds), The adequacy of the fossil record. 269–303. Wiley, New York.

Figure 5.19b: Courtesy of theedgeofforever2012.com.

Figure 5.19d: Courtesy of University of Maryland CATT Laboratory, www.cattlab.umd.edu.

Figure 5.20: Courtesy of Eat Seasonably Calendar © Behaviour Change Ltd. 2015.

Figure 5.21c: Courtesy of Edraw.

Figure 5.22a: Courtesy of Allen Hatcher.

Figure 5.23: From: Kutz, D. O. (2004). Examining the evolution and distribution of patent classifications. In Information Visualisation, 2004. IV 2004. Proceedings. Eighth International Conference on (pp. 983–988). Copyright © 2004 IEEE. Used with permission.

Figure 5.25b: Courtesy of Ivica Letunić, author of iTol http://itol.embl.de.

Figure 5.25d: Adapted from: Iliinsky, N. P. N. (2003). Book hierarchy. Unpublished.

Figure 5.26a: From: Juchem, C., Muller-Bierl, B., Schick, F., Logothetis, N. K., and Pfeuffer, J. (2006). Combined passive and active shimming for in vivo MR spectroscopy at high magnetic fields. Journal of magnetic resonance, 183(2), 278–289. Copyright © 2006 Elsevier. Used with permission.

Figure 5.26b: Courtesy of Martin Wattenberg.

Figure 5.28: From: Moody, J., and Mucha, P. J. (2013). Portrait of political party polarization. Network Science, 1(01), 119–121. Copyright © 2013 Cambridge University Press. Used with permission.

Figure 5.29: Courtesy of HyperPhysics.

Acknowledgments

We have a number of people to thank for making this work possible. First, thanks to David Ebert and Niklas Elmqvist for the opportunity to publish in this lecture series. Much thanks to Diane Cerra at Morgan and Claypool who did an excellent job managing and editing this book. We would like to extend our sincere thanks to both Niklas Elmqvist and Brian Fisher for their detailed and thoughtful reviews, which helped to improve the quality of the final product. We would also like to thank the members of the Insight Lab at Western University, with whom we had many discussions and brainstorming sessions that helped crystallize many of the ideas presented here. Last but not least, we would like to thank the Natural Sciences and Engineering Research Council of Canada without whose financial support this work would not have been possible.

Kamran Sedig and Paul Parsons

March 2016

CHAPTER 1

Introduction

1.1 OVERVIEW AND MOTIVATION

Much of the work in which people are engaged today involves extensive use of data and/or information.[1] We use information to make sense of and solve complex ecological problems, to predict and forecast natural disasters and weather conditions, to make evidence-based health and medical decisions, to identify and analyze trends in financial markets, to learn about mathematical and scientific phenomena, and even to plan and organize our social lives. The proliferation of information-based tasks and activities has led to increased interest in investigating the generation, storage, organization, use, design, communication, flow, ecology, and sociotechnical aspects of information. This investigation is taking place across many fields and disciplines including, among others, data/information/scientific visualization and visual analytics; information and graphics design; technical communication; library and information science; information systems; interaction design; instructional design; education; bio, health, and medical informatics; and human-computer interaction. Although these fields have distinct interests, some of their core concerns are similar. One common concern is how to effectively present relevant information to users.[2] A further concern lies in designing and implementing interaction mechanisms that allow users to work with the information—in other words, enabling a human-information interactive discourse—such that their tasks and activities are effectively supported.[3]

In the context of using computational tools, interaction with information is somewhat of a misnomer. Users cannot directly access the underlying data or information that is within the tool; rather, information must be *represented* to make it perceptually accessible and available for use and interaction. Information can be represented so that it is accessible to different perceptual modalities, such as hearing (auditory/sonic representation), touch (haptic representation), vision (visual representation), smell (olfactory representation), and even taste (gustatory representation). In this book, however, we focus specifically on *visual representations*, which are often referred to simply as *visualizations*. Unless otherwise noted, the terms *representation*, *visual representation*, and *visualization* are used synonymously throughout the remainder of the book.

---

[1] Throughout the book we mostly use the term *information*, and consider it to generally encompass *data*. The reader can think of them as interchangeable. Section 3.3 provides some analysis of these terms.

[2] The term *user* will be used generically throughout the book to refer to any viewer, analyst, reader, stakeholder, scientist, actor, librarian, student, learner, and any other person that uses data or information.

[3] We will use the terms *tasks* and *activities* interchangeably when it is not important to distinguish between them. They are examined in greater detail in Chapter 6.

In recent years, cognitive scientists have become increasingly aware of the important role of external representations (e.g., visual representations) in the tasks and activities that humans perform. There is now general agreement that external representations serve as fundamental components of cognitive functioning and behavior. Moreover, research has increasingly suggested that the interaction between internal and external representations is responsible for much of human intelligent behavior (Zhang and Norman, 1994; Hutchins, 1995; Kirsh, 2010). For designers or researchers concerned with the use, communication, and design of—or interaction with—information, the manner in which information is represented requires careful consideration. Much research relevant to visual representation use and design has been conducted over the past couple of decades. This research comes from multiple domains including, but not limited to, data and information visualization, visual perception, cognitive psychology, statistics, semiotics, cognitive engineering, education, diagrammatic reasoning, infographics, human-computer interaction, and user interface design. From this research, a body of generally accepted design principles, guidelines, and heuristics has been developed (see, e.g., Agrawala et al., 2011; Few, 2004, 2009; Mackinlay, 1986; Munzner, 2015; Tufte, 1983, 1990; Ware, 2012; Wilkinson, 1999). In addition to uncovering and developing principles and guidelines for design, numerous visualization types and techniques have been developed during this time (e.g., treemaps, parallel coordinates, streamgraphs, and heatmaps). These have joined a body of older, well-known representational types and techniques—such as scatterplots, bar graphs, and Venn diagrams—to constitute the space of known and established visualization techniques.

While a number of attempts have been made to categorize, taxonomize, and catalog visual representation forms and techniques in various ways (e.g., Aigner et al., 2011; Chi, 2000; Engelhardt, 2002; Harris, 1999; Heer et al., 2010; Li et al., 2015; Lima, 2011, 2014; Lohse et al., 1994; Meirelles, 2013; Parsons and Sedig, 2013a; Tory and Möller, 2004), it remains very difficult to capture all of them, as there are perhaps infinite ways of modifying and building on existing forms and techniques. Moreover, although such work is useful for organizing the design space, a catalog or taxonomy of techniques does not necessarily help designers generate visual representations in novel and creative ways. For example, in Meirelles (2013), visual representations are classified into the following six "structures": hierarchical structures: trees; relational structures: networks; temporal structures: timelines and flows; spatial structures: maps; spatio-temporal structures; and textual structures. While nicely classifying existing representations and describing some of their properties, these structures do not easily help designers understand what the basic, abstract "building blocks" are that can be used to generate the classified representations. Rather, designers have to begin the design process with the structures themselves in mind, which already suggests concrete forms and can make it difficult to think freely about generating new visual representations (Chapter 2 discusses this issue in more detail). Similarly, the other classifications, mentioned above, are directed more toward organizing the space of existing visualizations than toward helping designers systematically generate novel ones. In addition, the aforementioned contributions are primarily concerned with data/information visualization, and are not general enough to guide visualization design in other areas, such as education, knowledge management, business, and communication.

A common strategy among many designers is to simply use existing, familiar techniques, or to develop variations of them to fit their contextual needs. This strategy is appropriate in straight-forward situations where, for example, a bar chart or scatterplot adequately represents the under-lying data. In more complicated situations, however, this strategy may actually hinder creativity in the design process. Design frameworks, if well crafted, can alleviate such difficulties by guiding the thinking processes of the designer while still allowing for flexibility and creativity, and can also encourage reflective thinking about the design process itself. While there are a number of existing frameworks that provide support for thinking about different aspects of visualization design (e.g., Ainsworth, 2006; Burkhard, 2005; Card et al., 1999; Cheng, 2002; Chi, 2000; Hegarty, 2011; Javed and Elmqvist, 2012; Kosslyn, 2006; MacEachren, 1995; Munzner, 2009; Narayanan and Hegarty, 2002; Parsons and Sedig, 2013b; Tory and Möller, 2004; Wilkinson, 1999), their function is not to help designers "start from scratch"—without any visualizations or techniques already in mind—to generate novel visualizations for information of all kinds and from various domains.[4] By providing a conceptual design framework that starts from abstract, general patterns of organizing informa-tion, this book attempts to address the aforementioned issue. Designers should be able to use this book to generate new visualizations for all types of information without relying solely on existing solutions to guide their thinking.

Multiple authors have written about the notion of a "science of visualization," or a "science of visual representation" over the past few decades (e.g., Cleveland and McGill, 1984; Rensink, 2013; Thomas and Cook, 2005; Ware, 2012). Exactly what kind of science this would be remains unclear. However, an expectation that such a science would provide a systematic, coherent way of thinking about visualization use and design is clear and undisputed. In addition, that such a science should abstract from specific tools, techniques, and domains in order to focus on general principles, theo-ries, and patterns is inherently necessary. We anticipate that this book will make a contribution to a general "science of visualization," in the sense that it presents a framework that supports systematic thinking about the design of visual representations in a manner that is general and not tied to any specific contexts. We hope that it will encourage a more systematic approach to the study and de-sign of visual representations, and will inspire further discussion about what constitutes a science of visualization. In Chapter 2 we discuss our perspective on science and design in more detail.

---

[4] By "starting from scratch," we mean more than just providing support for choosing visual encodings. In all but simple cases, visualizations are more than arbitrary collections of encodings or visual marks. Designers should be able to start from more abstract patterns of organizing information, and subsequently deal with more concrete issues such as visual encoding.

While static visualizations are effective in supporting relatively simple human tasks, their efficacy wanes considerably when they are used to support complex tasks and activities, no matter how well they are designed (Dix and Ellis, 1998; Elmqvist et al., 2011; Kirsh, 2013; Morey et al., 2001; Pike et al., 2009; Sedig et al., 2003; Sedig and Morey, 2002; Sedig and Parsons, 2013; Tominski, 2015; Yi et al., 2007). Providing interaction mechanisms so that representations can be adjusted to suit different users, tasks, levels of expertise, and other contextual factors, can increase their utility and better support complex tasks (Parsons and Sedig, 2013b). In addition, there is considerable evidence to suggest that viewing multiple representations of the same underlying information can be beneficial in developing an accurate mental model of the phenomenon under investigation (Ainsworth, 2008; Baldonado et al., 2000; Kozma, 2003; Larkin and Simon, 1987; Roberts, 2007; Stenning and Oberlander, 1995; Sedig et al., 2005a). Interaction mechanisms can allow users to view and work with multiple representations, and to convert them from one form or structure into another. Furthermore, researchers have demonstrated that interaction with representations can extend and enhance thinking processes in fundamental ways (e.g., Kirsh and Maglio, 1994; Sedig et al., 2001; Liang and Sedig, 2010b). As cognitive processes are intrinsically temporal and dynamic, making representations interactive can potentially create a harmony and a tight coupling between them and the internal cognitive processes of the user (Brey, 2005; Clark, 1998; Kirsh, 1997; 2005; Sedig and Sumner, 2006). While simple tasks can be performed perceptually (e.g., identifying an outlier in a set of dots), more complex tasks must be performed cognitively, requiring sustained, effortful cognitive processing of the relevant information. In such cases, interaction mechanisms can be essential to the effective prosecution of the task, as actions that a user performs can enhance cognitive processes in a number of different ways (see Sedig and Parsons, 2013). Although this book can be used to design static representations for simple tasks, its real utility can be found when designing visualizations with interactive possibilities in mind—in other words, designing to support a human-information interactive discourse. This is especially true in contexts where interactive visualizations need to be designed to support complex activities with large and complex information spaces. This will be discussed more in Chapter 6.

In this book we present a framework for visualization design. The framework is made up of a number of elements that can function as conceptual tools during a design activity. The core of the framework is a pattern language—consisting of a set of 14 basic, abstract patterns—and a simple syntax for describing how the patterns are blended. We also present a design process, made up of four main stages, for creating static or interactive visualizations. The patterns that make up the core of our pattern language are basic, and can be thought of as letters of an alphabet. Using this metaphor, the pattern language describes how the combining and blending of these letters can result in the creation of words and sentences (i.e., visualizations). The 4-stage design process places the patterns at the core of the thinking of the designer, and employs a number of conceptual tools that help the designer think systematically about creating visualizations based on the information they intend to represent. In this process, the patterns—i.e., the letters—are used to generate visualizations—i.e., words, sentences, and paragraphs—to communicate information to the user. We will use this metaphor at various points throughout the book to explain and emphasize certain ideas.

1.2   AUDIENCE

The scope of this book is broad and may consequently be of interest to researchers, practitioners, and students in different fields. These fields include data/information visualization, visual analytics, scientific visualization, knowledge visualization, interaction design, human-computer interaction, digital humanities, information graphics, data/information science, biomedical and health informatics, digital libraries, education, and data journalism, among others. Although these aforementioned disciplines may have slightly different foci, their concerns often overlap insofar as they pertain to the design of visual interfaces, representations, and information presentation. Other disciplines that also share some of these concerns, such as technical communication and instructional design, may also find this book useful.

This book is intended for those who are interested in generating novel visualizations. This covers a wide range of potential readers who come from different disciplines and have different backgrounds and skillsets. Some may be programmers who are interested in visualization design; some may be graphic designers who are interested in visualization design; some may be journalists who are interested in designing visualizations for news stories; and so on. *This book is suitable for all such readers.* Readers of this book are not required to have expertise in design, visualization, or other related fields—we have attempted to make the material accessible to those with all levels of expertise. The only exception to this is that we sometimes make reference to well-known visualization techniques (e.g., treemap, Venn diagram, Sankey diagram) without describing them in detail. Readers who are not familiar with these techniques can easily look them up elsewhere. Chapters 2–4 are intended to provide a general introduction and overview of the state of the field, notable contributions, and fundamental concepts. While novices would likely benefit from doing some external reading, these chapters should provide enough general background to prepare readers for the remainder of the book. Finally, we expect that this book can be used by advanced researchers and practitioners, as well as by students. It could certainly be used as a component of a university course at all levels.

1.3   APPROACH, SCOPE, AND INTENDED USE

In this book, we approach the issue of visualization design at a general, conceptual level. This approach is different from many existing contributions that aim to give specific design prescriptions and/or technical guidance. Such guidance is usually focused on how to work with specific types of data, how to use visualization libraries or toolkits, and other implementation-related issues. This is not our goal. This book also does not comprehensively address perceptual and cognitive issues, user tasks and activities, interaction patterns and techniques, or specific visualization techniques. Although we do discuss these issues to an extent, they are beyond the scope of that with which we are concerned here; for those, there are plenty of existing resources to which readers can refer. One of the main motivators for writing this book is that much of the existing work has focused on these aforementioned issues, and not many researchers have attempted to deal with more general, conceptual and theoretical issues of visualization design. This is especially true when it come to issues related to design thinking—i.e., the cognitive activities in which designers engage, and the conceptual tools that are used while engaging in a design activity. In this book, we focus largely on conceptual tools and design thinking in the context of interactive visualization. This book is intended to support and promote systematic, coherent design thinking at an abstract level. In other words, this book should provide a conceptual framework that can help to organize a designer's thoughts and lead to principled design decisions. This will be discussed further in the following chapters.

In this book we are concerned with the process of designing *visual representations*. In the literature, the term *visualization* is used with many different underlying meanings. For example, sometimes it is used to refer to a whole tool or application; sometimes it used to refer to the process of visual encoding (i.e., visualizing), including data cleaning, transformation, and so on; and sometimes it is used to refer to the visual form (i.e., visual representation) at the visual interface. Here, we do not discuss issues pertaining to gathering data, cleaning it, mining it, and so on.

1.4   STRUCTURE

In Chapter 2 we provide relevant background information to situate the reader before presenting the visualization design framework. We present a brief overview of notable contributions to the field of visualization design. We also briefly compare our proposed framework to existing contributions. We revisit the notion of a science of visualization, discuss some differences between science and design, and briefly address the issue of design creativity in the context of such a science. We identify some issues with common terminology, examine the idea of design patterns, and discuss the role of frameworks and other forms of support in design. In Chapter 3 we introduce foundational concepts of the proposed framework. We introduce a number of conceptual "tools" that can help with systematic thinking about a number of issues related to visualization design. These include *systems theory* as a conceptual lens through which visualization design can be viewed; *information space* and *representation space* as spatial metaphors in design thinking; *encoding* and *representation*; *levels of abstraction*; *visualization techniques*; and *visual structures*, *marks*, and *variables*. These conceptual "tools" are intended to support design thinking in a general, coherent, systematic fashion. In Chapter 4 we present the 14 patterns of the pattern language. Each pattern is characterized and examined, and some simple examples of instantiations are given. In this chapter, we present only the building blocks (or letters, to use the metaphor from above) of the pattern language—i.e., the patterns themselves. In Chapter 5, we expand and develop further our pattern language, describe its simple syntax, and discuss how different patterns can be blended together to give rise to elaborate representations. In Chapter 6 we discuss relevant issues pertaining to human-information interaction. We examine activities, tasks, and interactions in the context of complex cognitive activities. We also briefly discuss interaction design, and present some considerations for designing interactive visualizations. In Chapter 7 we present the 4-stage design process. In Chapter 8 we present application of the design framework, where different examples are given to demonstrate the utility of different aspects of the framework in design situations. Finally, in Chapter 9 we briefly discuss and summarize the contributions of the book.

CHAPTER 2

Background

In this chapter we provide some relevant background information to situate the reader before presenting the visualization design framework. In Section 2.1 we provide a brief historical overview of the field of visualization design, identifying some key contributions and commenting on the need for more research in certain areas. In Section 2.2 we compare our proposed framework to existing contributions. In Section 2.3 we revisit the notion of a science of visualization. We discuss some differences between science and design, and also briefly address the issue of design creativity in the context of such a science. In Section 2.4 we identify some issues with common terminology, and also highlight some of the effects of language in design thinking. In Section 2.5 we present the idea of design patterns. Finally, in Section 2.6 we discuss the role of frameworks and other forms of support in design. Although this chapter provides the reader with some background knowledge, by no means does it constitute a comprehensive treatment of visualization- and design-related issues. However, it should suffice to prepare the reader for the framework itself, whose different elements are presented in the following chapters.

2.1 NOTABLE CONTRIBUTIONS

The work of the French cartographer Jacques Bertin seems to be the earliest attempt at developing a substantial theory of visualization design. His seminal work, *Semiology of Graphics* (1967/1983), was the most comprehensive in scope, and the most systematic in method, of any contemporary work on visual representations. In his book, he identified and classified a set of eight "visual variables" (size, value, texture, color, orientation, shape, and the two planar dimensions). Bertin considered these visual variables to be the fundamental units of visual communication and the basis of all forms of visual coding. Bertin's work has influenced numerous subsequent researchers, and his notion of visual variables has since been revisited, modified, and/or expanded by others (e.g., Mackinlay, 1986; MacEachren, 1995; Nowell, 1997; Carpendale, 2003; Heer and Bostock, 2010). Bertin emphasized the "strict" and "necessary" separation between the content (i.e., information or data) and the container (i.e., representational form). Subsequent research has demonstrated that the form of representations has at least as much influence on perception and cognition as does the content (e.g., Larkin and Simon, 1987; Zhang and Norman, 1994). Bertin's work influenced others (e.g., Lohse et al., 1994), who classified diagrams based on structural properties rather than content.

Subsequent key contributions were made primarily in the field of statistics by researchers such as John Tukey, Edward Tufte, and William Cleveland. These researchers recognized the potential for visual representations to facilitate analysis of statistical data. Tukey, for instance, observed that too much emphasis in statistics was placed on hypothesis testing (confirmatory data analysis), and that more emphasis should be placed on using data to suggest hypotheses to test (exploratory data analysis). Tukey emphasized the use of visualizations for this purpose. He invented the boxplot, and in his book, *Exploratory Data Analysis* (1977), he emphasized the utility of numerous other "graphical techniques" (e.g., histograms, scatterplots, and probability plots). Next, to promote the study and use of visual representations, was the statistician Edward Tufte. Tufte's first book, *The Visual Display of Quantitative Information* (1983), provided a practical theory of data graphics with a focus on quantitative statistical data. He developed principles for "graphical excellence" concerning clarity, precision, and efficiency, and for "graphical integrity" concerning context, proportionality, distortion, and others. He also took great strides to uncover much of the existing bad practice, and suggested guidelines for redesigning many common problematic design choices. His subsequent works (1990, 1997, 2006) have become broader in scope, expanding beyond statistical data to include all types of information. His focus has consistently been on principles for design of static representations (such as those concerning micro/macro readings, layering and separation, data-ink ratio), and on developing and promoting visualization techniques for static media (e.g., sparklines and small multiples).

Around the same time that Tufte's first book was published, Cleveland and McGill (Cleveland, 1985; Cleveland and McGill, 1984) observed the unscientific approach to the design of statistical representations of the time, stating "graph design for data analysis and presentation is largely unscientific" (Cleveland and McGill, 1984, p. 1). To establish "a few steps" toward a "scientific foundation" for representation design, Cleveland and McGill (1984) identified and described 10 elementary perceptual tasks that are used to extract quantitative information, such as direction, angle, volume, and shading, from visual representations. They performed a series of experiments to identify aspects of visualizations that helped or hindered accurate decoding of quantitative data (see Cleveland and McGill, 1984; 1986). Their work was generally concerned with methods for communicating statistical data (e.g., logarithms, residuals, distributions) with visualizations, and on the implications of representational features for perceptual tasks. While Cleveland and McGill focused specifically on quantitative information, Mackinlay (1986) extended their task ranking to account for non-quantitative information as well, many of whose visualizations lend themselves to different perceptual tasks. Mackinlay proposed a ranking of the accuracy of perceptual tasks with respect to encoding three different types of data—quantitative, ordinal, and categorical. As a result of this work, a number of design guidelines for matching variable properties of visual representations (also called visual variables) to different types of data were developed. Design considerations based on perceptual tasks will be discussed further in Section 3.7.

In the realm of cartography, MacEachren's book, *How Maps Work* (1995), provides a comprehensive account of a number of important issues in cartographic research and design. He draws from and integrates research from a variety of disciplines "to build an understanding of how maps are seen that can serve as a framework for research on and guidelines for map symbolization and design" (p. 147). His book combines research on perception and cognition with a semiotic approach to visual representations for cartography. He examines issues such as Gestalt grouping principles, selective attention theory, visual search models, perceptual categorization, and depth perception. MacEachren also provides a good discussion of semiotic concepts related to cartography, and builds on Bertin's visual variables.

There are a number of other notable contributions that are perhaps not fundamental, but are still highly relevant and useful. One is Harris's book, *Information Graphics* (1999), in which he identifies and characterizes over 3,000 different visualization techniques. Although he does not attempt to identify fundamental patterns or principles, his book still serves as a valuable reference for information graphics. With respect to visual perception and visualization design, Ware's books (2008, 2012) are probably the most comprehensive references. He covers a considerable number of issues in perceptual psychology, and also provides many guidelines and principles for visualization design based on an understanding of visual perception. For example, his books address, among others, the following issues: visual narrative, depth perception, color theory, mental imagery, visual pathways, motion, affordances, 3D objects, visual semiotics, visual salience, and Gestalt laws. Engelhardt (2002, 2006) has developed a framework for analyzing the syntax and meaning of various visualizations. His focus is on the "internal structure of graphics" and their syntactic principles. He integrates several aspects of visualizations into his framework, including structural and semiotic considerations. Wilkinson's book, *The Grammar of Graphics* (1999), presented a new, object-oriented way of thinking about quantitative, statistical graphics. He uses a grammar metaphor to provide a way of thinking about statistical graphics, including a discussion of their syntax and semantics. His book also provides a set of grammatical rules for creating visualizations of quantitative information. Table 2.1 provides a brief summary of the aforementioned contributions.

Table 2.1: List of notable research contributions, in order of publication year	
Publication	**Contribution**
Bertin (1967/1983)	One of the earliest attempts to provide a systematic, theoretical basis for visualizations
Tukey (1977)	Founder of exploratory data analysis using visualizations
Tufte (1983, 1990, 1997, 2006)	Proposed a practical theory for representing quantitative data and provided general principles and best practices for design and evaluation of visualizations
Cleveland and McGill (1984); Cleveland (1985)	Identified, suggested an ordering of, and conducted experiments pertaining to elementary perceptual tasks performed during extraction of quantitative information

Mackinlay (1986)	Extended work of Cleveland and McGill by adding rankings of visual variables for non-quantitative information
Lohse et al. (1994)	Performed a study investigating how people classify visualizations into hierarchically structured categories, and proposed a structural classification of visualizations
MacEachren (1995)	Combined research regarding perception and cognition with a semiotic approach to visualization for cartography
Harris (1999)	Developed a comprehensive taxonomy of information graphics for operational purposes
Wilkinson (1999)	Provided a formal framework for developing statistical graphics using an object-oriented grammar
Engelhardt (2002, 2006)	Identified various syntactic principles of visualizations and proposed a set of "building" blocks for constructing graphic spaces and objects
Ware (2008, 2012)	Developed a comprehensive collection of design considerations for visual perception

2.1.1   BRIEF COMMENTARY

During the latter half of the 20th century, many people did not take visualizations seriously as tools for effective communication of information. For example, visualizations were not thought of as useful for serious data analysis, and there was a prevailing assumption that they were "mainly devices for showing the obvious to the ignorant" (Tufte, 1983, p. 53). The work of early researchers such as Bertin, Tukey, Tufte, and Cleveland made great strides in legitimizing the use of visual representations for serious scientific work. Moreover, their work has served as a valuable foundation for much of the subsequent research in information visualization, information graphics, information design, and other related fields. That being said, although the aforementioned contributions are indeed valuable, they are not enough to form a coherent theoretical bases for a mature science of visualizations. In seeming agreement, after commenting on such aforementioned research, Thomas and Cook (2005) posit that "Although these design spaces and taxonomies are very promising, we are far from having a complete, formally developed theory of visual representations" (p. 71).

A number of areas that require further work may be identified. The first is in regard to the current terminology—e.g., terms such as map, diagram, chart, and so on, are not very precise, and are often not used consistently in existing literature (see Section 2.4 for more discussion of this issue). Second, much of the focus has been on identifying and examining various phenomena (e.g., perceptual issues, encoding techniques, heuristics for best practice) and not on identifying underlying patterns and integrating them into more comprehensive and general theories and frameworks. Third, most of the early research in this area was concerned with static visualizations. Although many of the findings and developments are still applicable to dynamic and/or interactive visualizations, research that accounts for novel features of interactive media is required. For instance, consider prescriptions from Tufte such as showing the viewer "the greatest number of ideas, in the shortest time, using the least amount of ink, in the smallest space" (Tufte, 1983); "enriching the density of data displays [is one of] the essential tasks of information design" (Tufte, 1990, p. 33); and "visual displays rich with data are ... frequently optimal ... the more relevant information within eyespan, the better" (ibid., 50). Certainly this is valuable advice for visualization design with static media. With interactive media, however, such prescriptive guidelines may be less than ideal and may even be inappropriate. For example, in the context of performing a complex activity with an interactive visualization tool, it may be best to initially encode only a small subset of the data, and allow the user to gradually explore more of the space at his or her own pace. This is especially true in the context of working with big data and large information spaces. Furthermore, depending on the stage of a user's overall activity, different degrees of density of the same visual representation may be appropriate. In other words, contrary to Tufte's advice, the highest density is not always optimal; flexible, human-centered tools should give users the ability to adjust the density and other properties of visualizations to suit their needs and preferences (see Parsons and Sedig, 2013b). Fourth, most early research was focused on supporting relatively simple tasks, such as identifying trends and outliers in a dataset, and was not performed in the context of today's sophisticated interactive interfaces that can support complex cognitive tasks and multi-layered human-information discourse. With today's interactive tools, analysts, for example, are not simply encoding datasets to find trends and outliers; rather, they are often engaged in activities that require a complex discourse with the underlying information—processing it, performing analytical operations, and carrying out other tasks that involve a sustained, coordinated partnership with the tool. Thus, as the aforementioned points demonstrate, the work of these early researchers can form part of the foundation of a science of visual representations, but other complementary work that addresses the above issues is required.

2.2   COMPARISON OF THIS BOOK TO EXISTING WORK

Some previous contributions to visualization design have focused on devising a language or a grammar of visualizations. For example, Wilkinson (1999) has developed what he calls *The Grammar of Graphics*. In this seminal work, he presents an object-oriented way of thinking about graphics, a language for describing graphics, and a set of rules for generating graphics. The scope of the work is not general, however, and is limited to quantitative graphics. He first defines a *graph* as a set of points—in the geometrical sense, being theoretical and having no dimensionality—and then defines a *graphic* as a physical representation of a graph. Based on these definitions, visualization techniques can be described using the main concepts of his language—e.g., graph, graphic, frame, variable, point, line, shape, and scale. For example, a scatterplot can be described as a "point graphic embedded in a frame," and a bar chart can be described as an "interval graphic bound to an aggregation function embedded in a frame." While constituting an influential foundation for creating quantitative graphics, and influencing the development of other work such as the visualization grammar Vega (Vega, 2013) and the plotting software ggplot2 (Wickham, 2009), his work is not easily generalizable to non-quantitative contexts. Furthermore, as his grammar is intended to be used by statisticians, computer scientists, and others interested in statistics and quantitative data, it is not intended to support visualization design that is non-quantitative in nature. While a statistician may find Wilkinson's work very helpful in supporting visualization design for statistical datasets, a science educator who wishes to design visualizations for biology education would not likely find the work to be of much help.

A somewhat similar but much broader approach was taken by Engelhardt (2002). In his doctoral thesis, he investigated what he called *The Language of Graphics*. In this work, a *graphic representation* is defined broadly as a "visible artifact…created in order to express information." Engelhardt analyzes graphic syntax in a general sense, providing a syntactic decomposition of graphic objects and an examination of syntactic structures. Furthermore, he presents a classification of 16 graphical types, including map, diagram, picture, statistical chart, link diagram, table, written text, and others.

There are a few essential differences between such previous contributions and our proposed framework. First, our framework is general, dealing with all types of visualizations and all types of information. Second, we are interested in supporting design thinking in the context of mapping information to visual representations. Previous contributions are often more focused on the visualizations themselves—including the rules for generating them and deconstructing them—and not on the conceptual tools that can be used by designers to systematically analyze an information space and think about how to visually represent the information. Third, our framework deals with very basic patterns for mapping information. As a result, it does not deal with design rules or prescriptions. At the level of very basic patterns, such as the ones we are proposing, it does not make sense to provide design rules or prescriptions (this will be discussed in more detail in later chapters). Fourth, we do not attempt to classify existing visualizations, as others (e.g., Harris, 1999; Lohse et al., 1994; Engelhardt, 2002) have done. We are focused more on helping designers generate new visualizations rather than on classifying existing ones. Fifth, we are fundamentally concerned with supporting design in interactive contexts. Previous contributions that focus on visualizations themselves, rather than the process of creating visualizations, often ignore the epistemic role and integration of interaction in visualization design. Finally, the sixth main difference is that our framework is not primarily intended to help designers create simple visualizations of single datasets—e.g., creating a scatterplot of a dataset containing the weight and height measurements of a group of people. Existing contributions already support this type of design. Our framework is most useful in helping designers systematically conceptualize non-simple (e.g., large, complex, dynamic, heterogeneous) information spaces, think about the tasks and activities of users, generate a multiplicity of relevant and potentially complex visualizations, and integrate a variety of interactions into their final visualization designs.

It is worth noting that, although the field of visualization design lacks a grand unifying theory, there is a well-established consensus on how designers can construct visualizations by mapping data points to visual primitives, as evidenced by well-known software tools and libraries such as D3 (Bostock et al., 2011), Lyra (Satyanaran and Heer, 2014), Protovis (Bostock and Heer, 2009), Tableau (Stolte et al., 2002), and ggplot2 (Wickham, 2009). Although such contributions are highly useful for implementation, and have helped to make visualization accessible to a wide audience, they are not conceptual frameworks that support thinking about visualization design in a comprehensive, coherent manner. Satyanaran and Heer (2014) note that systems making default decisions that limit control over visual design parameters, such as ggplot2 and Tableau, do not provide good support for developing novel visualizations. We are interested here in supporting creative design thinking—in helping designers generate novel visualizations in a systematic way. Our framework has no pre-defined palette of chart types; rather, it is intended to support the creation of a limitless number of novel visualizations. Furthermore, we are concerned with situations in which a designer has to choose how information should be visualized and presented, and not with situations in which the designer is simply implementing something that is already specified. Our framework has not been devised to compete with existing contributions that serve specific functions—such as describing a language for generating statistical graphics—nor is it meant to directly support implementation, as the aforementioned software tools and libraries do. Designers can use our framework to think about and plan their visualization designs, and then use a software library such as D3 to implement them.

As the reader progresses through the book, the differences between our work and others should become apparent. However, to summarize here, we can say that there is a gap in the field when it comes to supporting visualization design in a general manner—i.e., visualization design that transcends specific domains and specific types of data and information. This includes a lack of conceptual frameworks that can support systematic thinking about visualization design in a general manner, especially in complex contexts where interactive visualizations are needed to support users' tasks and activities. We have devised this framework to help bridge this gap and to stimulate further research in this area.

2.3   SCIENCE OF VISUALIZATION

There is a long history in the field of design research concerning the role of logic and method in design (see Alexander, 1964, Cross, 1981, 2011; Gedenryd, 1998; Stolterman, 2008). During the middle of the 20th century, there was a concentrated movement to make design more "scientific." Influenced by the obvious successes of the modern scientific method, especially as it led to rapid advances in engineering disciplines, design researchers increasingly searched for a design method that was similarly rational and based on objective grounds. While interest lasted for a small number of years, it later gave way to a more dichotomous attitude about the nature of science and design, resulting in a general consensus that a design method akin to the scientific method simply would not work (Gedenryd, 1998). Soon after, many original proponents of the movement completely rejected its underlying methodology, asserting that design is fundamentally a different enterprise from that of science. As design theorist Nigel Cross notes, designers and scientists have radically different goals: "unlike the scientist, who searches for many cases to substantiate a rule, and then one case to falsify it, the designer can be gratified in being able to produce just one satisfactory case that gives an appropriate result" (Cross, 2011, p. 28). Or, as Nelson and Stolterman (2012) note, the design process moves from universal, general, and even particular, toward the ultimate particular—a specific, contextually fit design outcome. This is in opposition to the scientific process, which moves toward the general, abstract, and universal. Asserting a fundamental distinction between the methods of science and design does not, however, remove the possibility of incorporating a scientific (i.e., systematic, disciplined, and consistent) attitude into the design process.

As mentioned briefly in Chapter 1, a number of researchers have proposed and/or discussed the notion of a science of visualization. While we do not attempt to conclusively deal with this issue, we briefly address it here. The term *science* is commonly used to refer to a body of accumulated information (or facts, knowledge, hypotheses, theories, and so on). In addition, it is commonly used to refer to a process, method, or way of thinking. Indeed, many eminent scientists have emphasized the latter meaning over the former (e.g., Sagan, 1990; Einstein, 1954; Bohm and Peat, 1987). We previously stated that this book can contribute to, and encourage more investigation and discussion of, a science of visualization. In this sense, we are thinking of the term *science* primarily in its second meaning—that is, a systematic, coherent way of thinking about something—in this case, visualization design. Thus we are not suggesting something akin to a natural science, nor are we naïvely proposing a scientific method of design, which has been so strongly rejected by most design researchers. Rather, we are providing a conceptual framework that is meant to enable a more *scientific approach* to the study and design of visualizations. We attempt to address fundamental aspects of how we can visually organize and represent information to support systematic, principled design. This book simply constitutes one possible contribution to such a science, as there are many other issues to be addressed that are not dealt with in this book. We thus anticipate that this book can serve as a component of a comprehensive, coherent "science" of visualization that will likely form more fully in the future.

Although the focus here is on a systematic approach to the design of visualizations, it should not be assumed that design creativity is precluded. A set of design patterns, such as the one being proposed here, acts as a conceptual framework that can give some structure to the design process, while still allowing for creativity at the level of implementation. In other words, a set of patterns provides a framework upon which any design can be anchored, but the patterns do not determine the specifics of the design. On the one hand, the constraints the patterns impose eliminate a large number of possibilities, but, on the other, still allow an infinite number of detailed designs. The narrowing of possibilities is, after all, an essential part of a practical design method (Salingaros, 2000). It is a common assumption that creative thinking occurs solely in the head. However, recent research in cognitive science indicates otherwise. Studies have suggested that external artifacts are in fact fundamental components of creative thought (see Kirsh, 2014). Design frameworks are one such type of external artifact that can support creative thinking in design.

Research shows that expert designers tend to start with abstract conceptualizations—rather than concrete solutions—to solve design problems (Cross, 2004). In addition, studies in psychology clearly show that a number of cognitive biases affect memory and decision making—an issue that is certainly a factor in making design decisions. For example, the availability heuristic (Tversky and Kahneman, 1973)—a type of mental shortcut that favors items that are more easily recalled from memory—can lead designers to call to mind techniques that are most well-known or recently encountered. This is not inherently problematic, but can be a limiting factor for creativity and innovation in the design of visualizations. It is very difficult for designers to start with a "blank slate" and not be influenced by previously encountered designs. Indeed, developing design ideas based on previous solutions—even if the solutions were very successful—can inevitably bias thinking and constrain it to the existing known solution space (Kirsh, 2014). With our framework, we intend to help designers generate novel visualizations by starting from abstract conceptualizations, and gradually moving to concrete solutions—all the while allowing for creativity in multiple stages of the design process.

2.4   COMMON TERMINOLOGY

Every discipline has a set of terms that are used to label relevant concepts. Having a common and accurate terminology (i.e., a system of specialized terms) can facilitate clear thinking about the relationship between terms and the concepts that they represent (Cabré, 1999). This can, in turn, promote effective communication between researchers in a discipline (Salustri and Rogers, 2008). Visualization design is a relatively young field, and involves researchers from numerous disciplines, including, among others, graphic design, statistics, cognitive and perceptual psychology, information science, cartography, and computer science. As a result, the numerous terms that are used to refer to concepts, as well as their interpretations, are not consistent. Consider the definitions for some of the following commonly used words, taken from the Oxford English Dictionary:

- **Diagram**: "a simplified drawing showing the appearance, structure, or workings of something; a schematic representation"

- **Graph**: "a diagram showing the relation between variable quantities..."

- **Chart**: "a sheet of information in the form of a table, graph, or diagram"

- **Plot**: "a graph showing the relation between two variables...a diagram, chart, or map"

- **Map**: "a diagrammatic representation of an area of land or sea showing physical features, cities, roads, etc."

The circularity and general lack of accuracy these definitions exhibit can be a problem. These terms have their genesis in times when there was no need for precision and rigor surrounding the discussion of visualizations. Consider the etymology of a few of these words: chart comes from the Greek *khartēs*, meaning papyrus, or book; graph comes from the Greek *graphikos*, meaning "writing or drawing"; diagram comes from the Greek *diagramma*, meaning "marked out in lines"; map comes from the Latin *mappa mundi*, meaning "sheet of the world." While these terms are fine for everyday colloquial usage, for academic research and for design discourse, more accuracy should be demanded. It is worthy to note that since the use of these terms has become so widespread, constructing new definitions for them is unlikely to be successful.

One of the desirable characteristics of a set of terms for any discipline is an avoidance of—or at least a controlled use of—synonymy (Cabré, 2003). Based on the definitions given above, the existence of synonymy is an evident problem. Consider the two visualizations displayed in Figure 2.1. Each one of these is referred to as a diagram (from L to R: Venn diagram, Sankey diagram). We can ask the following question: what does the term *diagram* say about these visualizations—about their similar characteristics and how they organize and communicate information? The answer is unsurprisingly: nothing substantive. Similar problems are encountered with the other terms mentioned above. Although most people do not see those terms as completely synonymous—e.g., maps typically represent relations of things in some (often geographic) space; plots often employ axes and represent relationships between variables in a dataset; and so on—their distinguishing features are often somewhat ambiguous. Perhaps the definition and everyday usage of these terms cannot be changed—which is not likely problematic. It is in the context of visualization design research and discourse where this synonymy can become a problem. We need to consider the terms that we use while thinking about how to generate a visualization. After all, the language that we use profoundly affects the way we think (Boroditsky, 2001; Levinson, 2003). As we will discuss in the next section, the use of patterns in design can help to alleviate this problem by providing a shared vocabulary and supporting abstract thinking. Our patterns are not meant to replace names used for common visualizations. Rather, they are meant to function as well-characterized, unambiguous terms that can be used in design thinking and communication. When used in the conceptual phase of design, they become a handful of "tools for thought" that can bootstrap the design process, enable creativity, and serve a generative function in design thinking, without relying on terms that may engender preconceived notions.

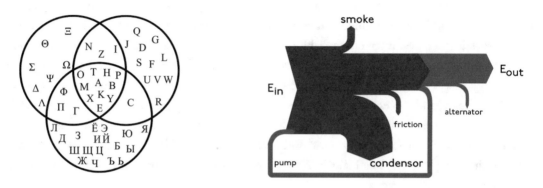

Figure 2.1: Two visualizations that are called diagrams: Venn diagram (L) and Sankey diagram (R). Sankey Diagram Courtesy of LHOON, commons.wikimedia.org/wiki/File:Sankeysteam.png.

2.5   DESIGN PATTERNS

One method of promoting systematicity and consistency in design, which has seen success in a number of domains, is the development and use of design patterns. The concept of a *design pattern* was made popular by the architect Christopher Alexander (see Alexander 1964; Alexander et al., 1977). His ideas have since been extended from architecture to other domains, such as software engineering (e.g., Gamma et al., 1995; Schmidt, 1995), human-computer interaction (e.g., Borchers, 2001; Tidwell, 2005; Sedig and Parsons, 2013), and information visualization (e.g., Chen, 2004; Elmqvist et al., 2011; Heer and Agrawala, 2006; Javed and Elmqvist, 2012). In recognizing the value of design patterns, some researchers have highlighted the need for their development to aid in designing visualizations. For example, in the seminal research agenda on visual analytics, Thomas and Cook posit that what is further needed is to "investigate whether we can *develop a set of visual design patterns* ... from which developers could draw to build new visualizations" (p. 73, italics added). In addition, Elmqvist et al. (2011) have suggested that there is a "space and a need for defining this kind of pattern language (or at least a pattern catalogue) for information visualization..." (p. 338).

General benefits of using patterns in design include: supplying a common language, which can improve communication in design teams (Schmidt, 1995) and can supply a *lingua franca* to facilitate discussion with those who are not specialists (Bayle et al., 1998); bridging many levels of abstraction (Winn and Calder, 2002); and, helping people abstract from implementation-level specifics to focus on general commonalities (Schmidt, 1995). In the context of visualization design, Elmqvist and Yi (2015) note that patterns—by capturing tested experience—can act as scaffolds for non-experts, effectively bootstrapping their design and evaluation activities. Although referred to as *design* patterns, patterns can also be useful for evaluation. Lanzilotti et al. (2011) conducted a study that compared pattern-based evaluation of e-learning systems to heuristic evaluation and user testing. The authors determined that using patterns for evaluation helped to reduce reliance on individual skills, increased inter-rater reliability and output standardization, permitted the discovery of a larger set of problems, and decreased evaluation costs. In addition, they noted that the common terminology that a pattern language promotes can make evaluation reports more consistent and easier to compare, and can enforce standardization and uniformity in evaluation. Elmqvist and Yi (2015) have proposed a set of patterns for evaluating visualizations. They note that it is especially difficult to evaluate visualization systems, and that one way to address this difficulty is to make use of evaluation patterns. They identify 20 patterns, which are general and reusable solutions to commonly occurring problems.

Although the utility of design patterns is widely acknowledged, not all patterns are equally as strong or as valid. In the context of human-computer interaction patterns, Dearden and Finlay (2006) have suggested that a weakness in many design patterns is their dependence on "particular and current user interface paradigm[s]," and that patterns should "*embody a timeless quality*, presenting a solution that is applicable regardless of particular platform or current technology" (p. 18, italics added). In order for patterns to be useful, they must be presented at a meaningful level of abstraction; patterns that are too abstract are of no use, and patterns that are too concrete do not generalize well beyond specific contexts (Dearden and Finlay, 2006). In addition, because they are not tied too closely to particular phenomena, patterns at basic levels of abstraction are more robust over time (Bayle et al., 1998).

2.6   DESIGN FRAMEWORKS

All design activity is mediated by the mind of a designer. Consequently, any design is influenced by the mental structures (schemas, representations, models, concepts) that the designer has. The outcome of a design activity is the result of an interplay among a number of elements, including the mental structures of the designer, the characteristics of the problem at hand, and the tools at the designer's disposal. A design framework is one such tool, which can guide the design process and help to organize relevant concepts in the design space. Based on the work of a number of design researchers (e.g., Schön, 1983; Krippendorff, 2006), Stolterman (2008, p. 63) suggests that designers appreciate and are inclined to use four forms of design support: (i) precise and simple tools or techniques (e.g., prototypes); (ii) frameworks that do not prescribe but that support reflection and decision-making (e.g., design patterns); (iii) individual concepts that are intriguing and open for interpretation and reflection on how they can be used (e.g., affordance); and (iv) high-level theoretical and/or philosophical ideas and approaches that expand design thinking, but do not prescribe design action (e.g., human-centered design). Forms ii and iv are most relevant for the discussion here. Perhaps they are found to be useful because they do not impose rigid constraints on the design process by prescribing specific low-level actions and procedures. Rather, they can serve a generative role in design thinking, can act as "tools" to support thinking, can aid in reflection, and can help designers be creative. Let us examine some research in design studies to help shed light on this issue.

Multiple studies have demonstrated that designers tend to hold onto their original ideas even in the face of obvious major shortcomings, and even when made aware of the existence of other, better solutions (Ball et al., 1994; Cross, 2004; Rowe, 1987; Ullman et al., 1988). Because there is a tendency to be attached to original ideas, using a framework to support systematic, critical thinking about a design may help to disrupt or remove the attachment. Research shows that one characteristic of expert designers is metacognitive awareness and control during a design activity (Kavakli and Gero, 2002). In other words, good design should be conscious and reflective—a good designer should reflect on how the situation has been framed and what strategies are being employed, so that they are explicated and examined (Schön, 1983). Various studies have shown that successful, experienced designers are proactive in this regard (Cross, 2004). Considering that research demonstrates the need for reflection in design, it makes sense that designers tend to appreciate and use frameworks that can support reflection.

Another explanation for the appreciation of these forms of design support is related to cognitive limitations of designers. Studies have shown that if the cognitive cost of engaging in a systematic, principled design process becomes too high, designers may abandon the process and favor more opportunistic, ad hoc solutions (Guindon, 1990; Visser, 1990). Thus, while designers appreciate and use forms of design support, such support should not impose a high cognitive cost. One contributing factor seems to be the number of options or possibilities that are made available to the designer as part of the support (e.g., framework or catalog). Fricke (1996) found that if options are too few, designers can become fixated on concrete solutions. Alternatively, when the options are too many, designers spend too much time trying to organize the abstract space of possibilities (ibid.). Although no specific number can be given to satisfy all cases, some studies have suggested that a relatively limited number of items may be best (Cross, 2004).

This book is intended to support design according to forms ii, iii, and iv mentioned above. The contents of the book constitute a framework that can support reflective thinking and decision-making regarding visualization design. A number of concepts (e.g., patterns, blendings, systems theory), which are presented in novel ways, can also be used to support creative design thinking. Furthermore, the number of basic patterns in our framework—14 in total—that allow designers to create countless visualizations, are neither too few nor too many. Finally, a number of the sections of this book are concerned with theoretical and philosophical aspects of visualization design, and can help designers think about visualization design in new ways. It should be noted that the existence of any form of design support, no matter how good, does not guarantee success in design. There are many considerations that go into design in general, and visualization design in particular, besides what can be captured in a single concept, framework, or theory.

CHAPTER   3

Conceptual Elements of Framework

In this chapter we discuss a number of concepts, including systems, space, information space, representation space, encoding, representation, levels of abstraction, visual structures, visual marks, and visual variables. These concepts, along with other ones that will be discussed in future chapters, form the main conceptual elements of the visualization design framework presented in this book. In Section 3.1 we provide an overview of the concept of systems. To account for the multidisciplinary nature of visualization research, and to transcend the limitations of domain-specific terminology, we propose that a systematic, scientific study of visualizations would benefit from a systems approach. One of the general maxims of the systems approach is that it is useful to consider phenomena as systems in order to understand and gain insight into them. A strength of the systems approach is its ability to provide a trans-disciplinary lens through which concepts can be discussed and applied consistently in and across different contexts and disciplines. Thus researchers and practitioners working in the seemingly disparate areas of scientific visualization, information visualization, information graphics, business intelligence, and educational technologies, for example, could have conceptually logical and terminologically consistent discussions about visualization design. In Section 3.2 we discuss the role of space and spatial metaphor in design thinking. In Section 3.3 we examine the concept of information space, the relationship and distinction between data and information, and the systems approach as a lens for analyzing information spaces. In Section 3.4 we present the concept of representation space, and the multi-level nature of this space. In Section 3.5 we discuss the difference between the two concepts of encoding and representing, and why we use the term encoding to refer to mapping of information. In Section 3.6 we discuss the abstract and concrete nature of techniques, marks, and structures. In Section 3.7 we briefly discuss what is meant by "visualization technique" in this framework. Finally, in Sections 3.8, 3.9, and 3.10 we cover the concepts of visual marks, visual structures, and visual variables respectively.

3.1   SYSTEMS THEORY

To engage in a more systematic, scientific approach to the design and study of visualizations, we propose that the concept of a system be used as a conceptual tool. Perhaps the best place to begin an examination of the concept of a system is with general systems theory, which can be seen as the science of systems themselves (Laszlo, 1972). General systems theory—henceforth, systems theory or systems approach—is concerned with general properties of systems, regardless of form or domain of application. Although numerous definitions for system exist, there seems to be widespread agreement on its basic characteristics. Thus a system can be defined as: an organized whole composed of parts that generate emergent properties through their interrelationships.

Although the features of particular systems vary, at a general level all systems are composed of three essential things: entities, properties, and relationships. For instance, a solar system has entities (e.g., planets, stars, comets), which have properties (e.g., size, color, temperature), and the entities and properties have relationships (e.g., orbits-around, next-to). What constitutes an entity, property, or relation, however, depends on the level of granularity at which a system is viewed. For instance, a number of entities such as planets, comets, and a star may, through their interrelations, result in an organized whole that exhibits emergent properties—i.e., that of a solar system. At a finer level of granularity, however, such entities may themselves be viewed as systems. Stars, for example, are organized wholes that exhibit emergent properties as a result of relations (e.g., gravity, nuclear forces) between entities (e.g., molecules). Such is true at the atomic and subatomic levels as well. Thus, such systems—indeed, virtually all existing complex systems—are in fact multi-level systems of systems (Skyttner, 2005). Consequently, an accurate way of describing the relationships among levels within a system is necessary. Systems theory uses the terms *sub-system* and *super-system* to express these relationships. Such terms are relative, however, and depend upon the perspective from which a system is viewed. As Laszlo notes, "a system in one perspective is a sub-system in another" (p. 14). For example, a solar system is a sub-system when viewed from the perspective of a whole galaxy, but is a super-system when viewed from the perspective of a planet. Therefore, an entity may be considered as a sub-system, a super-system, or simply a system; properties characterize entities and relationships at any one of these levels; and, furthermore, relationships may exist within one level of a single system (e.g., between two planets), among the levels of a single system (e.g., a chemical element and a star), or among entities and properties of any level of multiple systems (e.g., a star and a distant galaxy). Figure 3.1 depicts the multi-level nature of systems in an abstract manner.

While all systems have the aforementioned general characteristics, particular classes of systems have their own additional characteristics. Skyttner (2005) suggests that systems can be classified as concrete, conceptual, abstract, and unperceivable. Concrete systems have the additional characteristics of space and time; conceptual systems have the additional characteristic of being composed of ideas; and so on. Furthermore, each class of system has sub-classes, each with their own characteristics. For example, concrete systems can be categorized into living and non-living systems. Living systems have the characteristics of metabolism, self-regulation, and others. Such categorical specification is also the case for all other types of systems. Although more specification necessitates particular rather than universal terms, systems theory can be very useful for general conceptualization and interdisciplinary discussion. Indeed, one of the foremost strengths of the systems approach is its universality—that *its concepts and models can be applied to all phenomena*. In other words, the same language and concepts can be applied to all phenomena that meet the criteria

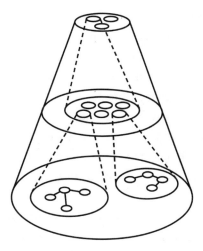

Figure 3.1: The multi-level nature of systems. Adapted from Skyttner (2005) © World Scientific. Used with permission.

of being a system—whether concrete, abstract, conceptual, or unperceivable (Skyttner, 2005). Thus, systems can be used as a powerful conceptual tool for design of visualizations—to describe the general properties and principles of the abstract (e.g., data/information) as well as the concrete (i.e., visualizations). Furthermore, by offering a consistent conceptualization and vocabulary, the systems approach can be used to discuss the relationships between the abstract and the concrete (e.g., the "mapping" of information to representational forms—discussed later in this chapter). More specific properties, principles, models, and frameworks may then be developed on top of the foundation of the systems theory depending on particular contexts or domains of application—domains such as business intelligence, medicine and healthcare, education, and others. One general property of systems that is useful here is whether they are open or closed. Open systems have exchange of information (input/output) with their external environment, whereas closed systems have their inputs determined once and receive no further information from the external environment. Because we are primarily concerned with interactive visualizations and tools, they can be considered as open systems that receive input from and give output to their users. Furthermore, information spaces can be open or closed. If information is coming into the space (e.g., data is being generated in real time), they can be considered as open systems. In contexts where the information does not change, they can be considered as closed systems.

3.2   SPACE

Visualization design is fundamentally concerned with encoding information—that is, converting information from one form to another. During the design process, information is converted from an abstract form (e.g., concepts, statistics, facts, data) to a concrete (i.e., visual) form.[5] Abstract information can be conceptualized as existing within a space, which can be called an *information space*. Visual representations can also be said to exist within a space—*a representation space*. Thus, design of representations is concerned with creating appropriate "mappings" from an information space to a representation space. While any instance of a representation space is literally spatial, an information space is metaphorically spatial. We suggest that there are a few reasons why using such spatial metaphor is useful in the context of visualization design. The concept of space naturally affords certain thinking styles—thinking in terms of containment and other geometrical concepts, for example. Spatial concepts, such as distance, location, area, and shape, are useful for thinking about any kind of information (Dade-Robertson, 2011). For instance, related concepts within a space can be thought of as being proximate, whereas unrelated concepts can be thought of as being distant. Research in cognitive science has demonstrated that human thinking is fundamentally grounded in spatial metaphor (see, e.g., Lakoff and Johnson, 1996). In fact, even subtle metaphors have a powerful influence on how we reason, conceptualize, solve problems, and perform other such tasks and activities (Thibodeau and Boroditsky, 2011).

Aside from the role of metaphor in general thinking, research has demonstrated that metaphor plays an important role in design thinking as well. For instance, metaphor is useful for the conceptual stage of a design process (Casakin, 2006), for structuring design problems (Gero, 2000), and for stimulating creativity in design thinking (Casakin, 2007; Snodgrass and Coyne, 1992). This may be especially true for abstract domains. In one study, Maglio and Matlock (1999) discovered that both novice and experienced users conceptualized the information that was underlying the visual representations they could see in terms of spatial metaphors (Maglio and Matlock, 1999). People tend to think of the information underlying representations as fundamentally spatial—in other words, as a space that contains items of information, and can be navigated along certain trajectories to access such information items (Dillon, 2000). This fact has implications for how we conceptualize and discuss data and information. For example, consider a context in which a tool must be developed to analyze and visualize a set of data that resides within a relational database. Although the designer is aware that the data exist within some columns and rows of the database tables, when it comes to the conceptual aspect of design, the designer is likely to think in spatial terms, especially when it comes to designing visualizations of the data. Indeed, researchers and designers seem to be naturally drawn to the concept of space. Many researchers interested in human-computer interaction, information visualization, visual analytics, information design, and other related areas often use the term "information space" to refer to the data and/or information that is visually represented (e.g., Allen, 1998; Benyon, 2005; Ahmed and Blustein, 2005; Thomas and Cook, 2005; Pirolli and Card, 2005; Boisot and Canals, 2007; Yi et al., 2007; Chen, 2010; Chen

---

[5] By the term *abstract*, we do not mean to suggest a distinction similar to information vs. scientific visualization. Rather, we are referring to information in its "non-visualized" form.

et al., 2009; Keim et al., 2010; Dörk et al., 2012; Heer and Shneiderman, 2012; Ware, 2012). While the term "information space" is rarely characterized in the literature, its prevalent use is a sign of the tendency to use spatial metaphor to conceptualize abstract domains. In the following two sections, we use the systems approach to characterize both information space and representation space, as well as to discuss the relationships between them.

3.3   INFORMATION SPACE

An information space is a body of information that is conceptualized as having spatial character-istics. In this book we use the term *information* in a broad sense: information refers to anything from which meaning can be construed—e.g., financial sales data, spreadsheets, databases, textual documents, web logs, medical records, videos, and images. Such information may exist at the time of design, or may be generated dynamically through, for example, statistical procedures, algorithms, and analytic processes. For instance, consider a visual analytics application in which medical data is initially provided; a set of statistical, analytical processes are performed while the tool is being used, and new data is derived as a result. Both the original and the new, derived data are part of the information space—along with any other information that is relevant in that particular context. Furthermore, such analytic processes may be continually occurring—making the information space dynamic, *with much of the information existing only potentially until it is generated*. All such infor-mation—whether original or derived—is contextually relevant and is thus contained within the information space. Within any given information space, information can be combined and blended from multiple sources and environments. Moreover, such information can be quite diverse—being structured and/or unstructured; homogeneous and/or heterogeneous; dynamic and/or static; large and/or small. For example, an information space may contain unstructured web logs, photographs, streams of incoming data, and/or static documents.

Many researchers interested in visualizations tend to refer only to *data* as the underlying con-tent that is being visualized. As was mentioned in the previous section, in most cases of design—at least where some degree of complexity is involved—we posit that the concept of *information space is far more powerful than just data in terms of supporting design thinking*, for two primary reasons. First, as described previously, the use of spatial metaphor is essential in human thinking in general, and has been shown to have benefits for design thinking in particular. Such is likely the reason why researchers tend to use the concept of information space without characterizing it. Second, data does not conceptually capture all of the possibilities for the underlying content of visualizations. As it is typically used, *data* refers to recorded discernable states of some phenomenon. For example, MRI data are recorded states of the structure or function of an organ under certain conditions; public health data are recorded states of some behavior (e.g., hand washing) within a population; financial data are recorded states of financial instruments (e.g., derivatives) and their differences over time. Such data are typically stored in databases, spreadsheets, and other documents and file systems. In application areas such as intelligence analysis, investigative journalism, and climate analysis, the concern is typically how to visualize and explore sets of data. In other application areas, however, such as education, business, software engineering, communication, and knowledge management, there may be no dataset (e.g., database, spreadsheet, table) to be visualized (e.g., see Ainsworth, 1999; Arcavi, 2003; Bodner and Domin, 2000; Burkhard, 2004; Gerjets et al., 2010; Hegarty and Kozhevnikov, 1999; Jong et al., 1998; Markman, 1999; Moreno et al., 2011; Peterson, 1996; Snyder, 2014; Thomas et al., 2012). In these areas, the concern is often how to visualize and explore concepts, objects, ideas, propositions, and processes—not necessarily a specific set of data. Furthermore, in many of these fields there are *model-driven* visualizations—i.e., visualizations whose features and behavior are not dictated simply by a designer encoding data points, but are dictated by a model, which is an abstract characterization of something. Although models are often developed by abstracting from data in order to capture general characteristics, designers may not be working with the original data from which a model was developed. Rather, they may be simply employing an existing model—e.g., a model that is embedded in a software library. We may want to differentiate these different types of design as *data-driven* and *non-data-driven* visualizations. Although in reality such a clear dichotomy is not always apparent, the point we are trying to make is that designers are not always working directly with datasets, simply attempting to encode data points in the best way possible. Design activities are often much more open-ended and complex, requiring the use of different conceptual tools to support various aspects of design thinking. In order for our framework to meet the goal of being general and comprehensive, we have attempted to ensure that the terminology we employ covers all contexts, whether visualizations are directly data-driven or not. Thus, we find it generally more comprehensive to discuss the "information" and/ or the "information space"—which can contain all kinds of data, concepts, ideas, processes, models, and so on. In situations where a designer has a set of data to visualize, the data and the information space can be thought of as being synonymous. Conceptually, however, the spatial metaphor may still be beneficial for supporting creativity and conceptualization during design. Generally speaking, then, "information space" is conceptually much broader than "data," and any given information space can contain a collection of different datasets, databases, images, videos, and all kinds of digital documents, as well as statistically derived or algorithmically generated data.

In this book, we propose that systems theory can act as a useful tool in conceptualizing and discussing information space in a general manner. As mentioned previously, systems can be of many types. An information space is a spatial system. All information spaces, besides the most simple, whether abstract (e.g., financial statistics) or concrete (e.g., a solar system), are composed of entities, properties, and relationships that exist at many levels of granularity.[6] In this book, a generic term—*information item*—is used to refer to any entities (including information sub-spaces), properties, and relationships within an information space. This includes terms such as "data point," "data element," or possibly "information element." However, as an information space is a multi-level spatial system, the term "information item" can refer to sub-systems (i.e., information sub-spaces), properties, or relationships at any level. For example, an information item can refer to a single "data point," a whole dataset, a set of datasets, an individual document, a paragraph within a document, an image, an idea, a set of ideas, and so on, depending on the context of use. In other cases, information item does not refer to any particular thing; rather, it is simply a term that is used during discussion and conceptualization to refer to something generic within an information space. To provide an example, consider a large and complex information space containing electronic health records. Depending on the context—i.e., the need of the designer during conceptualization, during discussion with other designers, while providing a design rationale for educating designers or presenting design ideas—"information item" may be used to refer to an individual patient, a property of an individual patient (e.g., blood pressure), a hospital, a region, a disease trend, or a radiology image. Thus, the term "information item" is used, when necessary, to refer to a generic thing within an information space, and, when necessary, to provide particular details.

While "information space" is a commonly used term in existing literature, its meaning is not consistent. Often no distinction is made between underlying information and the representation of that information, and thus both are simply referred to as "information space." Furthermore, information space is sometimes used to refer to the physical environment in which one is situated. In this book, however, "information space" does not refer to the physical environment, nor does it refer to the visualizations—rather, an information space is not directly accessible, and can only be accessed and interacted with through visualizations at the visually perceptible interface of a tool. Similarly, researchers sometimes conflate data with its visual representation, and refer to what is perceptible (e.g., visualizations) as "data." Such habits are less likely to lead to clear design thinking and dis-

---

[6] It is important to note that even though systems theory and entity-relationship modeling use similar terms (e.g., entity, relationships), they are fundamentally different. Entity-relationship modeling captures explicit relationships among entities and describes them. For example, given three entities "manager," "employee," and "social security number," we can model their relationships as: "manager" is-a "employee" and "employee" has-a "social security number," where is-a and has-a are relationship descriptors. These can be easily translated into an entity-relationship diagram, and can also be easily stored in a relational database. Systems theory provides a lens for conceptualizing a whole space of information (e.g., concepts, objects, data), including a way of thinking about multiple levels of systems (i.e., sub-systems and super-systems) and multi-level relationships. The language of systems theory provides a number of concepts and terms, such as emergence, coordination, equilibrium, holism, equifinality, complexity, openness/closedness, and entropy, that can act as powerful tools for conceptualizing entities, relationships, and properties in ways not possible with entity-relationship modeling. Thus, while there is some similarity between them, systems theory is a more general and robust conceptual tool than is entity-relationship modeling, and employs a much richer vocabulary for describing complex relationships and processes within information spaces.

cussion. In this book, the terms *data* and *information item* are used to refer to only abstract things within an information space, and not to things that are visually perceptible in representation space.

property
quantity
geolocation
measurement
datum
category
database record
image   = information item
document
set of documents
video
dataset
subset of an information space
federated datasets

…

Figure 3.2: Information item as a generic term referring to all possible things within an information space.

3.4   REPRESENTATION SPACE

Representation space is the space in which information items are encoded into perceptible visual forms. Although visualizations are displayed within a literal, physical space—usually on some sort of computer display—the concept of representation space is an abstraction, and is not tied to any particular platform or technological implementation. A representation space becomes reified in a physical display such as a typical computer display, tablet or other mobile device, projection display, or any other yet-to-be-developed display. Conceptually, however, representation space does not change. It is worthwhile to note that representation space contains visualizations not only of information items from an information space, but also of controls, navigational elements, and other information that a designer determines to be useful for a user. For example, Figure 3.3, shows an instance of representation space in the visualization software *Tableau*. Items from the information space are encoded (e.g., oil wells as circles) alongside controls and other elements (e.g., sliders, checkboxes) that help the user work with the encoded information. In this book, we are generally concerned with only visualizations of information items and not with other visualizations that may exist in representation space.

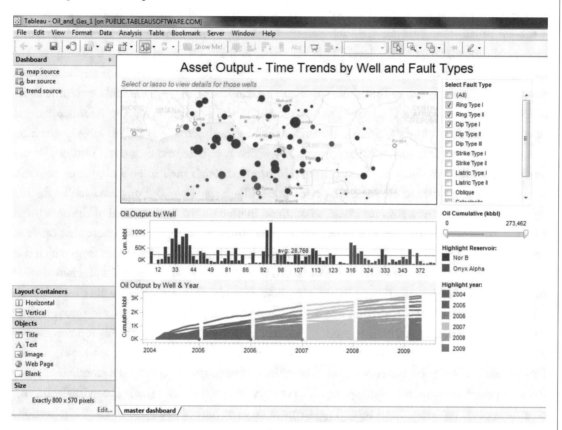

Figure 3.3: An instance of representation space in the software *Tableau*. Items from the information space are encoded alongside controls and other elements that help the user work with items in the information space. Courtesy of © 2016 Tableau Software.

From a general standpoint, representation space is composed entirely of visual representations. What is considered as a representation, however, depends on the perspective from which it is viewed. For example, consider a representation space that contains two scatterplots. The entire space can be considered to be one visualization; both scatterplots together can be considered as one single visual representation; each scatterplot can be considered as a separate visualization; the individual dots within a scatterplot can be considered as visualizations. In each case, the consideration is correct—that is, each thing being considered *is* a representation of something. However, such ambiguity and imprecision are not desirable characteristics of a scientific approach to any phenomenon. To provide some precision in this regard, similar to the case with information space, systems theory can be used as a powerful tool for conceptualizing and discussing representation spaces. A representation space is also a spatial system that is composed of entities, properties, and relationships. Furthermore, any visualization that is not simple and atomic can be decomposed into a sub-set of sub-visualizations (i.e., sub-systems), each of which can be decomposed into other sub-visualizations, down to the level of atomic visual entities, discussed later in this chapter. For instance, Figure 3.4 depicts one way in which an instance of a representation space can be analyzed through the lens of systems. The whole space can be considered as a system. This system is composed of sub-visualizations (i.e., sub-systems), which are the geographic map representations, one of which is represented by the circle at the top of the cone outlined in green. This sub-visualization is composed of multiple sub-visualizations, one of which is represented by a circle outlined in red on the second level of the cone. These sub-visualizations, through their interrelationships, result in the emergent properties of the overall representation. These sub-visualizations can each be further decomposed into sub-visualizations (e.g., as outlined in blue on the bottom level of the cone), and so on, down to the atomic level of indivisible visual entities. It is important to note that one can decompose the same representation in different ways, depending on the perspective from which it is being viewed. This is because systems theory is a conceptual lens, and, as described previously, what constitutes a system or sub-system in any situation depends on perspective. In this regard, such a lens for analysis enables clear thinking about what representations communicate through their emergent properties, in addition to how to interact with them—e.g., which sub-representations within a representation space should be the objects of user action (for more on this see Sedig and Parsons, 2013). Designers can consider visual entities and what they encode at the atomic level; how their properties and interrelationships result in systems at a higher level (i.e., super-visualizations); how such visual representations result in systems at an even higher level; and so on. Visualizations at higher levels of the overall system communicate increasingly more emergent properties. While the representation space shown in Figure 3.4 is relatively simple, because of the universal descriptive power of systems theory, such an analysis can be conducted on all representation spaces—of any size, composed of any number of representations, and with any degree of complexity. Consequently, using systems as a lens through which representations are viewed enables logical consistency, and provides a language that is generally applicable, making systems theory a helpful conceptual tool in the development of a science of visual representations.

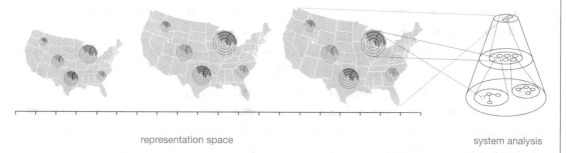

representation space   system analysis

Figure 3.4: A representation space viewed as a multi-level system.

3.5   ENCODING vs. REPRESENTING

The terms *encoding, mapping, visual mark*, and *representation*, along with their various word forms, are frequently used in the existing literature. Their meanings and relationships, however, are not always made very clear. Such concepts are of fundamental importance, and their definitions must be accurate and precise for them to be considered valid components of a science of visual representations. The first required clarification is that the word encoding can be used as both a noun and a verb. Its prefix, *en*, is used to form verbs that express conversion into a specified state or form. Therefore, information encoding refers to the conversion of information from one state or form to another—e.g., from information space to representation space. This conversion process is typically what is meant by information "mapping." In other words, to encode information is to "map" information items from an information space to a representation space. Encoding as a noun, on the other hand, refers to a perceptible visual item in a representation space that is the result of the mapping process. To avoid confusion, throughout this book we use "encoding" only as a verb, and use "visual mark" to refer to a visual item in the representation space. Visual marks in representation space are items such as dots, lines, and simple shapes, and are often the building blocks of more complex visualizations.

While encoding is concerned with conversion of information from one form to another, representing is concerned with signification. To represent means to stand for or to signify. A representation, then, is something that stands for or signifies something else—e.g., a visual form that stands for or signifies an information space or an information item. *A representation does not necessarily encode everything that it represents.* For example, a simple dot can represent (i.e., stand for) a person. It would not, however, encode the entirety of the information space of the person. In other words, it does not map all of its entities, relationships, and properties to a visual form in the representation space. Technically speaking, the dot can be considered a visual mark (i.e., an atomic visual item) as well as a visual representation (i.e., a visual form that stands for a set of information items); however, *what the dot encodes and what it represents are not equivalent.*

To summarize: to encode information is to perform a mapping from an information space to a representation space; visual marks are the objects of such mapping within a representation space; moreover, visualizations may be simple visual marks, but are usually composite forms that combine and integrate multiple visual marks in such a way that communicates emergent properties of information items; and visual representations stand for or represent information items, but do not necessarily encode all of their attendant data. Figure 3.5a demonstrates a situation in which an information space has a number of sub-spaces with different types of documents in them. The designer has used four visual marks (circles) to represent each sub-space, but the circles do not encode all information from the sub-spaces. Rather, they simply encode the existence of the document sub-spaces, their type (using color), and their size (using area). Figure 3.5b shows the same information space, where the designer is encoding information items from only one of the sub-spaces (i.e., rows from the table). None of the other information items within the other sub-spaces are encoded, although they exist within the information space. Figure 3.5c also shows the same information space in which an item (a document) from one of the sub-spaces is being represented such that the length of each paragraph is encoded using visual marks (bars). This example makes use of a number of variable properties of visual marks, which are also commonly referred to as visual variables. These are discussed in detail in the following section.

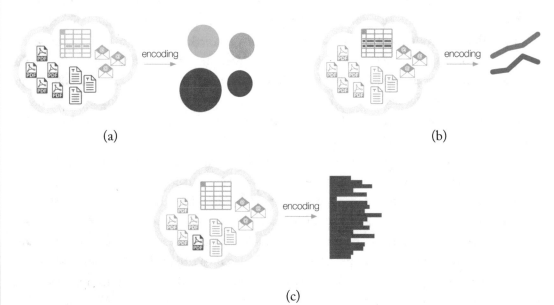

(a)   (b)

(c)

Figure 3.5: Different ways of mapping information items from information space to representation space. In each case, different items are encoded.

3.6   ABSTRACT vs. CONCRETE: TECHNIQUES, STRUCTURES, AND MARKS

In this chapter, we have started from the highly abstract notion of systems, and have used it as a lens to conceptualize information space, representation space, and the mapping between them—all of which are themselves abstractions. In the next three subsections, as we move toward lower levels of abstraction, we will discuss the concepts of *visualization technique*, *visual structure*, and *visual mark*. Even though each one of these concepts is not highly abstract, and is close to the level of concrete, physical instances, it can still be categorized along two lines: abstract and concrete. While subsequent sections will explain these concepts in more detail, we believe that it is important here to alert the reader to the key distinction between abstract and concrete techniques, structures, and patterns.

By abstract here we mean conceptual and non-physical, where geometric and physical details are not decided upon. By concrete we mean physical, where geometric and physical details are decided upon and are implemented in the representation space. For example, a visualization technique is a specific method or template for visualizing information in a specific context. An example is TileBars (Hearst, 1995), which is a technique used in search interfaces for showing the relationships between words in a search query and the documents that are returned. At this level of specificity, the technique is still an abstraction. Once a designer implements it, choosing specific physical properties (e.g., sizes, colors, shapes, and so on), it can be considered a concrete visualization technique. A similar categorization is true for visual structures. Structures, as we see them, are intermediary configurations that suggest some type of organization of information. They are typically derived from or based on a pattern or a blending of the patterns in our framework (this will be discussed more in Chapters 4 and 5). For example, a designer may select some patterns and blend them in a certain way, then decide that a cellular structure is needed in order to partition the information. However, this cellular structure is still abstract. At this level, the geometric and physical details are not specified—whether it is 2-dimensional or 3-dimensional, and whether its constituent cells are circular, hexagonal, rectangular, spherical, cuboid, and so on. Some types of abstract structures are cellular structures, coordinate structures, hierarchical structures, multi-dimensional glyph structures, containment structures, and geometric structures. Finally, the same categorization can be made for visual marks. Visual marks are atomic visual entities. A few types of visual marks are lines, dots, letters, and shapes. Even though being close to physical instances, at this level of characterization they are still abstractions. A designer can decide that a shape is needed, without specifying the exact type, dimensionality, size, color, and other physical properties. Not until the marks are implemented in the representation space, and all of their geometric and physical properties are decided upon, are they considered concrete.

We find that this distinction can help with clear design thinking. Even though a designer may be thinking about visual marks, for example, he or she can first decide on the abstract marks, and then implement them and determine their physical details. This type of distinction can also enable clear design discourse. A team of designers may discuss and decide upon a set of consistent visual marks at the abstract level, and then each designer may implement them differently in very specific contexts at the concrete level.

3.7   VISUALIZATION TECHNIQUES

In our framework we employ the concept of a *visualization technique*. While the term visualization technique is frequently used in the literature (sometimes written as graphical technique or method), there is no commonly agreed upon definition. Although visualization techniques can be thought of as patterns, they are not universal, and are instead confined to specific contexts of use (e.g., specific domains, specific data types, specific tasks). For example, LineSets (Alper et al., 2011) is a technique for visualizing sets of elements and their relationships. However, it is not used to visualize, for example, the evolution of species over time. In other words, it is confined to a specific context of use. As another example, TileBars (mentioned previously) is a technique used in search interfaces for showing the relationships between words in a search query and the documents that are returned. However, it is not used to visualize, for example, relationships in social networks. Thus, techniques can be thought of as methods or templates that can be used in specific visualization design contexts. Compared to the patterns in this framework, techniques are more specialized, closer to concrete visual forms, and can have a more prescriptive aspect to them. However, as mentioned in Section 3.6, techniques can still be abstract, without specific physical details. Once implemented in the representation space, they become concrete. Researchers and designers in different disciplines create their own techniques that are useful for their relevant data, concepts, users, tasks, and so on. For example, different techniques are used in statistics, mathematics, science, business, education, biology, chemistry, engineering, and other domains and disciplines. These techniques and others can be found in a host of references: Aigner et al., 2011; Börner, 2010, 2015; Card et al., 1999; Chi, 2000; Coopmans et al., 2014; Cuoco and Curcio, 2001; Eilam and Gilbert, 2014; Glasgow et al., 1995; Hansen and Johnson, 2005; Harris, 1999; Jonassen et al., 1993; Lima, 2011, 2014; MacEachren, 1995; Malcolm, 2004; Markman, 1999; Meirelles, 2013; Moktefi and Shin, 2013; Pauwels, 2006; Peterson, 1996; Spence, 2007; Tufte, 1983, 1990; Ward et al., 2015.

3.8   VISUAL MARKS

Visual marks are the primitive building blocks of visualizations. In other words, they are atomic visual entities. All complex visual representations can be broken down into basic visual marks. People sometimes refer to visual marks as *encodings*, however, as explained before, to avoid confusion, in this book we do not use the term *encoding* in this way.

There are various ways of classifying basic visual marks into general types. The most common approach is to classify them according to the number of dimensions that they require on the plane. Thus, there are four possibilities: points (zero dimensions), lines (one dimension), surfaces or areas (two dimensions), and volumes (three dimensions). This classification appears to have originated with Bertin (1967), and has been adopted and used by others since then (e.g., Card et al., 1999; Börner, 2015; Munzner, 2015). Although this is the most common classification, and can be useful in some ways, we have found that it can also be quite confusing, especially for new students and novice designers that are learning about visualization design. When presented with this classification, a number of questions tend to arise. For example, at what point does a point become a surface? How small should a circle be to be considered a point rather than a surface? In other words, it appears to designers that distinguishing between the different classes is somewhat of an arbitrary process. The problem actually lies in the fact that the language of mathematics (geometry) was used to develop this classification, and the terms (as often used in visualization design) are not semantically equivalent to their theoretical geometric values. That is, the classes proposed by Bertin exist in an abstract, theoretical realm, but they do not apply to the marks that are concrete and visible. In geometry, the use of the term *point* is meant to convey the idea of a unique location in Euclidean space. As a result, geometric points do not have any dimensional attributes (e.g., length, area, volume). In the visualization literature, however, *point* is often used to refer to a concrete visual mark, and thus it certainly does have an area. As Bertin states, "a point represents a location on the plane that has no theoretical length or area…by way of contrast, the mark which renders it visible can vary in size, value, texture, color, orientation, and shape…" (Bertin, 1967/1983, p. 44). It is very important to notice here the distinction between the theoretical, geometric type (point, line, area) and the mark that "renders it visible." This seems to be the main point of confusion for most people. The situation for line and area is similar to that of the point. For example, a line has length but no area; however, "…the mark which renders it visible can vary according to all the variables other than those involving position on the plane: in width, value, texture…" (Bertin, 1967/1983, p. 44).

Bertin was a cartographer, and it made sense that he was fundamentally concerned with the plane. A basic concern of cartographers lies in mapping items from an information space to a representation space such that the planar dimensions of the representation space convey spatial dimensions in the real world. Thus it makes sense to assume theoretical geometric primitives in the conceptual framework for cartographic visualization design. Outside the realm of cartography, however, and in the realm of general visualization design, there are many cases in which such planar considerations are not important. It is not difficult to imagine situations in which multiple visual representations are present in a representation space, but their planar attributes have no real meaning, except to convey their existence. In other words, we do not suggest that such representations have no planar dimensions; rather, we suggest that their point, length, or area on the plane do not encode any information.

As the distinction between theoretical geometric primitives and concrete visual marks can be confusing for designers, we do not see any reason to promote it here for general visualization design. However, it can still be useful to think of points, lines, areas, and volumes as types of marks. Some examples of concrete visual marks are shown in Figure 3.6.

Figure 3.6: Examples of visual marks.

One classification that can be useful in design is based on the binary classification of *analogical* and *symbolic* representation. By analogical we mean simply that the representation visually resembles that which it is representing, and can be interpreted without knowledge of convention. By symbolic we mean that the representation does not visually resemble that which it is representing, and can be interpreted only by understanding its convention. Figure 3.7 provides a very simple example. If we want to encode the existence of an information item, such as a car, there are two basic types of visual marks that can be used: (1) a symbolic visual mark, such as a dot, in which the user must be provided with rules for decoding the semantics of the mark; or (2) an analogical mark in which the user can decode the semantics of the mark without any additional information. Either visual mark can then be modified to encode additional information about the car other than its existence. Section 3.10 deals with this issue (i.e., using visual variables) in greater detail.

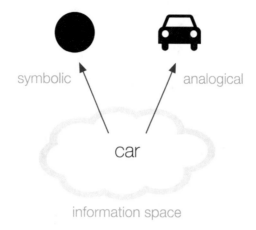

Figure 3.7: A visual mark that represents an information item (car) can be symbolic or analogical.

3.9   VISUAL STRUCTURES

Visual structures are configurations that suggest some type of organization of information. As discussed in Section 3.6, concrete structures have physical details and are implemented in the representation space. They are composite, being made up of two or more visual marks—e.g., multiple lines, dots, or shapes. Compared to abstract structures, concrete structures have more specificity. For example, a designer may decide in abstract that a coordinate structure is needed. One example of a concrete coordinate structure with more specificity is a three-dimensional Cartesian coordinate system with specific physical properties in the representation space. Another example is a two-dimensional polar coordinate structure also with specific physical properties. In other words, once a designer chooses an abstract structure, there is still variability in how the structure is implemented in a concrete form. Examples of abstract structures are coordinate structures, cellular structures, hierarchical structures, tabular structures, containment structures, and geometric structures. Figure 3.8 presents some examples of typical concrete structures.

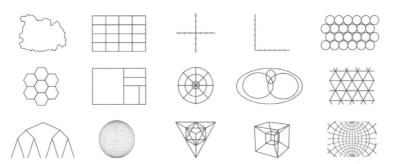

Figure 3.8: Examples of visual structures.

3.10   VISUAL VARIABLES

All visualizations—whether visual marks or more complex structures—have a number of visual properties that can meaningfully encode information. These properties are variable, and are thus often referred to as *visual variables*.[7] Specifically, any visual representation can vary with respect to a number of properties, the more common ones being: planar dimensions (usually two, but three is possible with 3D projection displays), color, size, angle/orientation, texture, shape, curvature, and motion (the last property pertains to only dynamic displays). Some of these general properties have more specific sub-features: color can vary with respect to hue, saturation, and luminance; the planar/ spatial dimensions can vary with respect to alignment (aligned/unaligned) and depth (3D position); size varies depending on the geometric dimensionality of the representation, and motion can vary with respect to its pattern (e.g., regular oscillation, random jumping). The other four properties do not have sub-features that are used for encoding purposes. Table 3.1 lists these general properties (i.e., visual variables) and, if applicable, their variable sub-features.

In Figure 3.9 very simple visual marks are used to demonstrate how visual variables can be used to encode information. Figure 3.9a shows three visual marks (circles). They encode the *existence* of three items in an information space, but none of their variable properties (e.g., color, shape, size, orientation) are being used to encode other information. In other words, all we know from Figure 3.9a is that there are three items—the designer has not encoded anything else from the information space. Figures 3.9b to f, however, demonstrate how different visual variables can be used to encode more information: size (b), color hue (c), shape (d), spatial location (e), and color satura- tion (f). In most design situations, multiple variables are used together to encode information, as in Figure 3.9f, which uses the size, spatial location, and color-saturation variables. These variables can

---

[7] Terms such as *visual channel*, *retinal variable*, *graphical property*, and others are often used in the literature synon- ymously with *visual variable*.

be used to encode more information from the information space—in addition to just the existence of items—such as size, class, value, importance, and others.

Table 3.1: Some general visual variables and their specific variable features	
Visual Variable	**Variable Sub-features**
Color	saturation, luminance, hue
Spatial	aligned planar, unaligned planar, depth
Size	length (1D), area (2D), volume (3D)
Motion	pattern, direction, velocity
Angle/Orientation	--
Curvature	--
Shape	--
Texture	--

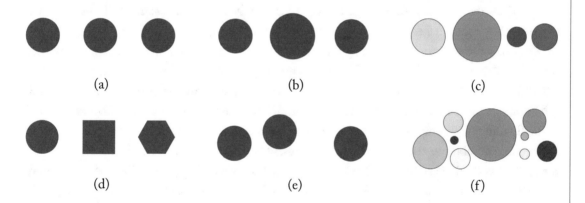

(a)   (b)   (c)

(d)   (e)   (f)

Figure 3.9: Visual variables of visual marks (a), whose properties can be adjusted to encode information items: using size (b); using color hue (c); using shape (d); using spatial location (e); using color saturation (f).

The example above uses very simple visual marks for the purpose of illustrating the concept of visual variables. In most real design scenarios, more complex visual representations are involved. Figure 3.10 illustrates similar ideas to the one above in a more realistic context. In Figure 3.10a, visual marks (dots) are used to encode the existence of counties in the U.S. Ignoring the underlying representation (the geographical map) for now, the only variable property of the marks that is being used to encode data is the spatial position, which encodes latitude and longitude (i.e., location) of the counties. In Figure 3.10b, the designer has used the size variable to encode the population of each county. In Figure 3.10c, the designer has used the additional variable of color hue to encode a property—e.g., has used two hues to encode two possible values for a property of the counties, such as whether the majority of residents in the country prefer one sport over another. While this example is still simple, the purpose is to place the discussion of visual variables in a realistic design context.

(a)   (b)   (c)

Figure 3.10: A visualization that uses different visual variables to encode information about counties in the U.S.

While the previous examples introduced visual variables in the context of simple visual marks, designers can make use of visual variables with more complex visualizations as well. Because we view the representation space through the lens of systems theory, designers can view any system, sub-system, or super-system within the representation space as a visual representation. For example, in Figure 3.11a, we can consider the whole representation as a visualization; or the node-link sub-system can be considered as a visualization; or each bar can be considered as a visualization; and so on. By thinking in a multi-level systems fashion, designers can think systematically about how to apply visual variables to even large, complex representations at many different levels. For instance, the same visualization can be thought of as a sub-visualization of a bigger visualization, as in Figure 3.11b, and visual variables can be used at the level of the sub-visualizations. Similarly, that whole visualization can be considered as a sub-visualization of an even bigger visualization, as in Figure 3.11c, where visual variables are again applied to the sub-visualizations. The result is that designers can think about using visual variables to encode information at many levels of the representation space, from the highest level down to the level of visual marks, in a systematic and coherent fashion.

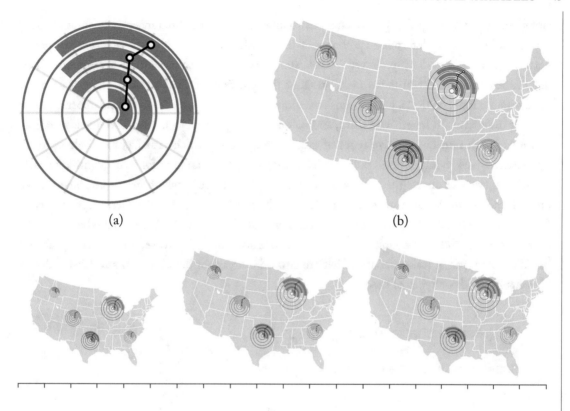

(a)

(b)

(c)

Figure 3.11: Applying visual variables at different levels of visualizations.

As a designer, it is important to know about different visual variables. However, it turns out that not all visual variables are perceived with the same degree of accuracy by the human visual system. More specifically, some variables are more effective than others in supporting certain kinds of perceptual judgments and comparisons. These perceptual judgments and comparisons are often referred to as *perceptual tasks* or *visual tasks*. Visual tasks can be roughly divided into low-level and high-level tasks. Low-level tasks occur mostly in the pre-attentive stage of visual processing. However, as we do not wish to get into the minutiae of pre-attentive versus attentive processing, we simply use the terms low- and high-level. Low-level visual tasks refer to goal-driven behaviors that are performed by the human visual system while looking at visualizations (e.g., perceiving the comparative sizes of two visual marks), and not to higher-level goal-driven behavior (e.g., categorizing a set of items while solving a problem).

One of the first studies to look at the effects of visual variables on low-level visual tasks was done by Cleveland and McGill (1984). Cleveland and McGill were concerned specifically with quantitative tasks that people perform when looking at visualizations—e.g., judging position along a common scale, judging differences in volume, and judging differences in orientation. They conducted a study to determine the order of these tasks on the basis of how accurately people perform them. Figure 3.12 provides a visual summary of the results of their experiments. They determined that, for quantitative tasks, judging the position of visualizations along a common scale was the most accurate, judging position along non-aligned scales was second-most accurate, judging differences in length was third, and so on. It is important to note that the study was not concerned with non-quantitative types of tasks—e.g., judging which visualizations belong to the same category. Mackinlay (1986) built on this work to develop a ranking for non-quantitative tasks as well. Figure 3.13 shows the ranking that he devised for three types of tasks—i.e., quantitative, ordinal, and nominal. One can see that position is the most accurate for all types; length is second for quantitative but much lower for ordinal and nominal; and so on. We have grouped the variables that have common relative rankings using color to assist the reader to see how rankings change based on the types of tasks. Variables that are crossed out cannot be used with any reliable degree of accuracy for those types of tasks.

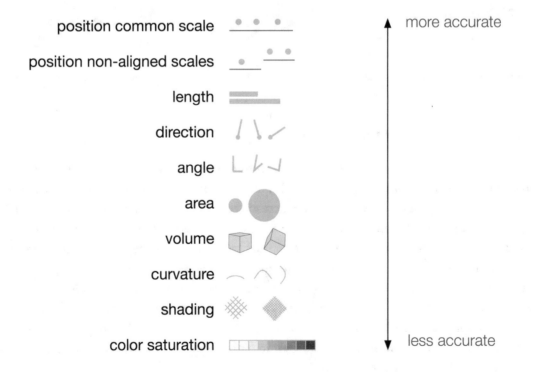

Figure 3.12: Ranking of visual variables based on accuracy of perceptual tasks requiring quantitative judgments. Adapted from Cleveland and McGill (1984).

It is important to keep in mind that the effects of visual variables on visual tasks occur primarily at the low level of visual processing. Furthermore, these effects are especially accentuated when the time available to perform the tasks is very short (Carswell, 1992). In Chapter 6 we will engage in a deeper examination of these different types of tasks (including interactive tasks), specifically in the context of human-information interaction and complex cognitive activities.

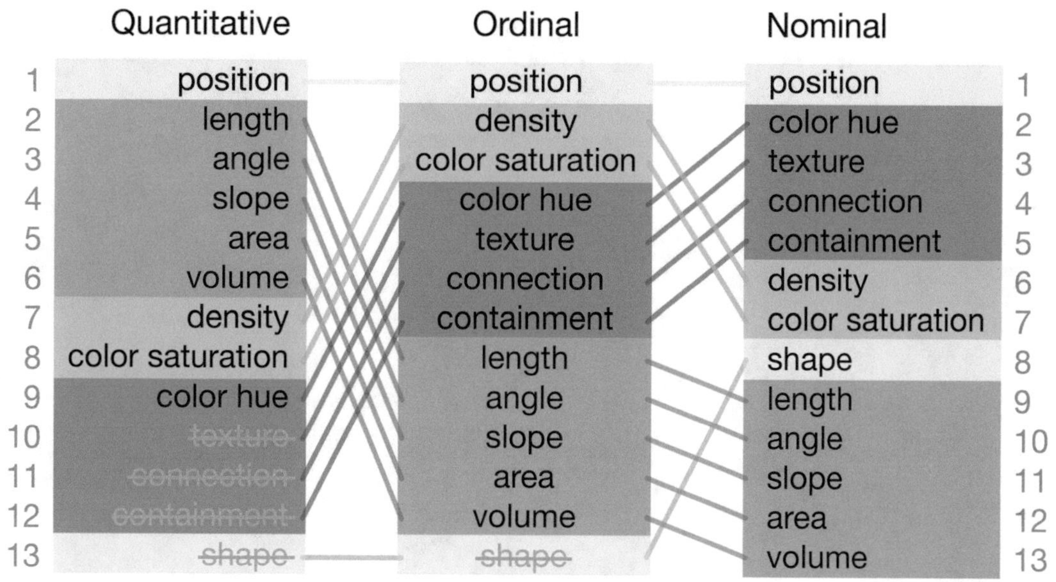

	Quantitative	Ordinal	Nominal	
1	position	position	position	1
2	length	density	color hue	2
3	angle	color saturation	texture	3
4	slope	color hue	connection	4
5	area	texture	containment	5
6	volume	connection	density	6
7	density	containment	color saturation	7
8	color saturation	length	shape	8
9	color hue	angle	length	9
10	texture	slope	angle	10
11	connection	area	slope	11
12	containment	volume	area	12
13	shape	shape	volume	13

Figure 3.13: Ranking of visual variables based on accuracy of perceptual tasks requiring quantitative, ordinal, and nominal judgments. Adapted from Mackinlay (1986).

CHAPTER 4

Patterns

In this chapter we present a set of patterns—i.e., the building blocks or the letters—of the pattern language, with the language itself being the core element of the visualization design framework. In Chapter 5, we will expand and develop further our pattern language, describe its simple syntax, and discuss how different patterns can be blended together to give rise to elaborate representations. The patterns presented in this chapter, when used together, can inspire the creation of innumerable techniques and visual representations—from simple to very complex visualizations. This chapter is organized as follows. In Section 4.1 we define what we mean by the term *pattern* and how we have mined and identified our patterns. In Section 4.2 we provide a list of the names of the patterns and the rationale for choosing the names. In Section 4.3 we discuss the purpose and application of patterns in design thinking. In Section 4.4 we briefly explain why we do not have anti-patterns for our patterns. In Section 4.5 we explain the concepts of instantiation and blending of the patterns. Section 4.6 forms the bulk of this chapter. In this section we characterize each pattern, mention its usefulness in design thinking, and provide some simple examples. In Section 4.7 we categorize the patterns into three groups and provide a summary. Finally, in Section 4.8 we describe how patterns in this book should be used.

4.1 DEFINITION AND IDENTIFICATION ON PATTERNS

The term *pattern* is used in various contexts with many different meanings and connotations. Here we borrow from Salingaros (1999) and define a pattern as "a regularity in some dimension." Patterns are derived empirically from observations, rather than from first principles (Salingaros, 2000; Dearden and Finlay, 2006). In their comprehensive review of patterns in human-computer interaction, Dearden and Finaly (2006) discuss the process of identifying patterns. They recount Alexander's (1979) process, in which he finds places that exhibit what he calls "the quality without a name," and then tries to identify the distinguishing characteristics that account for the success of the selected design solution. He then seeks to identify key "invariants" that are common to all good designs and practice. Although identifying good practice is important for developing patterns, it can be considered the "least part of the achievement" (Fincher, 1999, p. 338). Bayle et al. (1998) note that it is relatively easy to observe phenomena which can be put into a pattern-like form, but it is much more difficult to use these observations to develop and explicate good patterns. "Practice can be captured at any scale, but it is the combination of capture and abstraction that makes the presentation of the ideas coherent" (Fincher, 1999, p. 339). *In order to be useful, patterns must present abstractions at a meaningful level of granularity.*

One commonly used method of identifying patterns is the process of "pattern mining"—extracting patterns from observing previous designs. This process is commonly used in many fields, including software design, architectural design, and interaction design (e.g., Gabriel, 1996; Meszaros, 1996; Iacob, 2011; Sedig and Parsons, 2013). In the process of identifying the patterns in this chapter, we have examined thousands of existing visualizations. These can be found in numerous books and articles (e.g., Cairo, 2012; Tufte, 1983; Lohse et al., 1994; Heer et al., 2010; Harris, 1999; Mazza, 2009; Meirelles, 2013; Spence, 2007; Aigner et al., 2011; Lima, 2011), and existing tools and techniques (e.g., Cytoscape, Shannon et al., 2003; NetLogo, Wilensky, 1999; Table Lens, Rao and Card, 1994; EdgeMaps, Dörk et al., 2012; Vizster, Heer and Boyd, 2005; GapMinder; DEMIST, van Labeke and Ainsworth, 2001; Polyvise, Morey and Sedig, 2004; Film Finder, Ahlberg and Shneiderman, 1994; HARVEST, Gotz et al., 2010; Tableau, Stolte et al., 2002; NodeTrix, Henry et al., 2007). In finding patterns for our pattern language (i.e., a core component of the visualization design framework), our intention and focus has been the identification of basic, abstract, fundamental patterns that are independent of any particular technology, platform, medium, or domain, that exist at a consistent level of abstraction. As was stated in Chapter 2, for patterns to be useful, they must be presented at a meaningful level of abstraction that keeps them robust over time and makes them "timeless"—not too abstract to render them difficult to use, and not too concrete for them not to generalize beyond specific contexts. In finding such patterns, we have searched for common, fundamental organizing patterns, similar to Alexander's "quality without a name." We have focused on identifying patterns that map information items from an information space to a representation space at a basic level of abstraction. In this process, we have regarded the patterns' communicative utility in concrete visualizations and their mapping utility in design situations as distinct concerns. This is because as patterns become blended and instantiated, they can be used to communicate many things.

We have identified *14 basic, abstract patterns*. Combinations of these patterns can be found in countless visualizations. However, it is important to note that these **patterns are abstractions and not visualizations**.

4.2   NAMING OF PATTERNS

In choosing names for each pattern, common terms, such as diagram, chart, map, graph, and so on, have been avoided. Recall that these terms have different connotations for different groups of people, and generally do not have any fully agreed-upon meanings (see Chapter 2 for more discussion of problems with current terminology). Accordingly, the pattern names that we have selected are very close to the dictionary definition of the concepts they represent. They are suggestive of the design structures that they inspire and how these structures organize information in the representation space. For example, one of the patterns we have identified and propose is named *List*. This term suggests placing items in a sequential, successive order; another pattern is named *Hierarchy*, which suggests multi-level organization and parent-child relations; and so on. The 14 abstract patterns that are the building blocks of the pattern language are: *Area, Branch, Cell, Coordinate, Cycle, Fusion, Group, Hierarchy, Link, List, Spectrum, Stack, Token,* and *Track*.

4.3   APPLICATION OF PATTERNS

It is also important to note and emphasize that the purpose of having the patterns in this book is to bridge the gap between the information space and the representation space in the context of design thinking and intention. In design thinking, their application is in *information mapping*—that is, designers can use them as thought supports to determine how to give organization to information items as they are mapped onto the representation space. As the number of visualization design possibilities is very high, thinking about the patterns and their blendings can give some high-level support to design thinking about how to engage in this mapping and organization. Figure 4.1 depicts this space and identifies the aspect of design intention with which we are concerned. We would like to emphasize again that the patterns do not characterize *concrete* visualizations in the representation space. Such visualizations are the already instantiated representations of pattern blendings—any given visualization is an *instantiation of one or more of the patterns* and communicates information in a visually perceptible form, and can never itself be a pattern. For instance, as will become clear later, a coordinate plot in the representation space is a concrete visualization, often an instance of a blending of the abstract patterns Coordinate and Token, or Coordinate and Fusion.

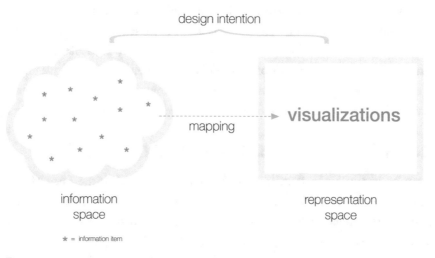

Figure 4.1: Design intention is concerned with mapping information items from an information space to a representation space and giving organization to visualizations within the representation space.

4.4   ANTI-PATTERNS

The basic concept of a pattern, as "regularity in some dimension," can be extended into the solution space of design problems. This is what Alexander and colleagues did in proposing their pattern language. In doing so, timeless archetypes can be captured—i.e., solutions to common problems, which can be used "a million times over, without ever doing it the same way twice" (Alexander et al., 1977). When extended into the solution space, design patterns are essentially reusable, proven solutions to common design problems. For such design patterns, common problems can be labeled, described, and prescriptions can be given as to how to overcome or manage the problem. To understand the visualization design framework in this book, it is critical to recognize that *the patterns here (in and of themselves) are not solutions to design problems*. Rather, our patterns are more basic than those presented as solutions to problems. To employ the language metaphor used in Chapter 1—patterns in the solution space involve identifying common problems and best practices in putting together (i.e., "designing") words, sentences, and paragraphs. At this level of design, best practices can be identified and captured as design patterns. Our patterns, however, are like the letters of the alphabet—they are more basic than the words, sentences, and paragraphs, and do not function as patterns in the solution space. Thus, our 14 patterns have no prescriptive element.

Since our patterns do not operate in the solution space and are non-prescriptive, they have no relevant anti-patterns. The term *anti-pattern* was coined by Andrew Koenig in the context of design patterns in software development (1995), and was later used in development of design patterns in other disciplines. Anti-patterns are patterns that appear to be effective and appropriate, but in actuality have adverse consequences. In this book, we do not present any anti-patterns. The reason for this is simple: as outlined before, our patterns are basic building blocks, not solutions to design problems. As such, it does not make sense to identify anti-patterns along with these types of patterns. To revisit our alphabet metaphor, it makes no sense to present anti-patterns along with letters of an alphabet. Since our framework does not promote the use any particular patterns, nor does it prescribe any particular order or forms of their blendings, it seems moot to have anti-patterns for the characterized patterns here. The patterns presented in this chapter are to support design thinking and analysis. They do not capture good or bad practices, or offer solutions to design problems.

4.5   INSTANTIATION AND BLENDING OF PATTERNS

A pattern can be instantiated—that is, it can become represented in the form of an instance or occurrence. An instance of a pattern is usually a visual structure—from a simple visual mark, such as a shape, to a more elaborate structure, such as a polar grid. Any pattern can be instantiated using many structures. As will be discussed in detail in the next chapter, patterns can also be blended, and these blendings can be instantiated to give rise to all types of visual structures. In the characterizations of the patterns in this chapter, we use two terms frequently: *structure* and *based*—as in a structure that is based on or derived from a certain pattern or blending of patterns. By using the term "structure," we are referring to the arrangement and functional or semantic relations between or among a set of visual representations or marks that together instantiate a pattern—thus bringing about a "structure." By using the term "based," we are referring to the fact that a pattern or blending of patterns is involved in giving rise to a structure. In this sense, a structure should not be thought of as a one-to-one mapping from a pattern to a specific visual entity within the representation space. In fact, many structures are based on a blending of multiple patterns, and in such cases it is not possible to ask which individual pattern a structure is derived from—the question itself is inaccurate. It is important to note that, in this chapter, we sometimes use examples that are generated from blending multiple patterns; we use them solely for demonstration purposes to depict a pattern under discussion.

Additionally, it should be noted that some of the 14 patterns allow for immediate derivation of visual structures (e.g., Token, Coordinate, and Area), while others only allow derivation based on blending with other patterns (e.g., List). To demonstrate the utility of some of the patterns in this chapter, we will use example visualizations that are based on simple blendings, usually involving a given pattern and the Token pattern. Many of the examples given in this chapter use abstract, simple visualizations that function as placeholders. At this point, we do not want to present any visualizations that are derived from elaborate blendings or are final and concrete, as they can distract from the main function of the examples, which is to suggest to the reader what simple instances of a pattern can look like. Plenty of such examples will be given in the next chapter where the syntax of the pattern language is presented.

Finally, as these 14 patterns are abstractions, there are no particular shapes or geometric forms for any of them. We may refer to general structures, such as circular shape or polar coordinate, as possible forms that patterns can take, without these being concrete instances. Particular shapes, colors, textures, and so on are considerations at the more concrete level of encoding rather than at the abstract level of patterns. In other words, the fact that a visualization is an instance of a blending of patterns does not say anything about the visualization's size, shape, texture, or color.

4.6   CHARACTERIZATION OF PATTERNS

Before we characterize the patters, a point needs to be explicated. In naming the example visualizations in this section, we use common terms such as *diagram, chart, map, graph,* and so on. We use these terms only because readers may be familiar with them, and using them will help us demonstrate the utility of the patterns by keeping consistent with current naming conventions. However, we emphasize that these terms have different connotations for different groups of people, and generally do not have any fully agreed upon meanings (see Chapter 2 for more discussion of problems with current terminology).

4.6.1   TOKEN

Designers can use this pattern to map items from an information space onto a single unitized visual representation. Token is the most used pattern for representing data, and, as such, most visual representations are based on a blending of Token with other patterns. This pattern is useful when a designer wants to represent an item, such as an object, property, value, system, operation, location, direction, feature, function, or process, in an individual, discrete fashion in the representation space. In some cases, an item might be very intricate, large, and/or complex within the information space, yet the designer wants to map it onto a simple, individual visualization in the representation space; in such cases, the item is being represented but its features, properties, and relationships (i.e., other items) are not all being encoded (see Section 3.4 for more discussion of this issue). The simplest instance of Token can be seen when a designer maps an information item onto an atomic visual mark. Examples of this form of instantiation are a dot representing a city; a color representing the percentage value of a variable; and a simple symbol representing a digit, a letter, or an operation. While Token is an elemental pattern, it is also very broad, as there are many ways of instantiating this pattern. Designers can map information items onto atomic visual marks; they can also create composite visualizations in which a number of visual marks are put together in a unitized fashion. In such cases, although the representations are not atomic and are composed of multiple, discrete parts, the designer is organizing them in a unitized, holistic manner. Examples of this form of instantiation are a multidimensional icon representing a type of house and its features; a word representing a concept; a stick figure representing a human being; a Chernoff face representing characteristics of baseball managers; and a composite, 3D shape representing a geometric structure. In addition to the above types of instantiations, an image, when representing an individual, singular information item, can be regarded as an instance of Token. Figure 4.2 shows a number of instances of the Token pattern.

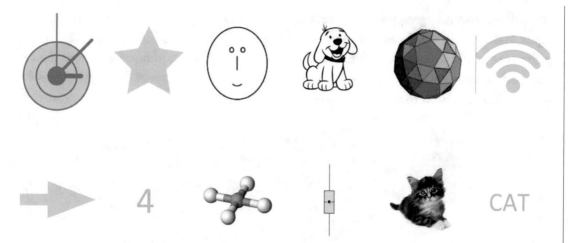

Figure 4.2: Examples of Token-based visual representations.

4.6.2 AREA

Designers can use this pattern to map items from an information space onto visualizations such that their boundary, shape, region, and/or area are encoded in the representation space. The Area pattern can be useful when a designer wants to represent information items that have two or three spatial dimensions with a definite shape, boundary, or region in the representation space. Figure 4.3 shows some structures that can be derived from this pattern. This pattern is often blended with others to create different types of geographic and geometric visualizations, such as statistical maps, dot distribution maps, proportional symbol maps, cartographic maps, and geometric shapes.

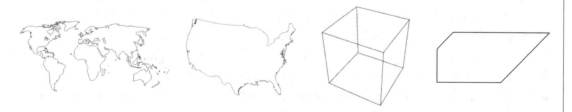

Figure 4.3: Examples of Area-based structures.

4.6.3 BRANCH

Designers can use this pattern to map items from an information space onto visualizations and organize them in the representation space in a branched and/or subdivided fashion. This pattern can be useful when a designer wants to represent converging and/or diverging (e.g., temporal, categorical, conceptual, or geographic) features of information. In addition, this pattern allows the designer to show whether information items originate from or terminate in other sources. The Branch pattern can be blended with others to produce common visualization techniques, such as Sankey diagrams, parallel sets charts, flow charts/diagrams, cladograms, and alluvial diagrams. Figure 4.4 shows a structure that is based on this pattern.

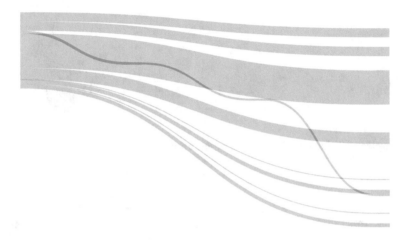

Figure 4.4: Example of a Branch-based structure.

4.6.4 CELL

When using this pattern, designers map information items onto visualizations and organize them by segmenting, compartmentalizing, or containing them within cell-like structures. This pattern can be useful when designers want to create partitions or compartments for a set of visual representations according to their temporal, ordinal, nominal, categorical, conceptual, or spatial properties, and want to represent distributed membership, containment, division, or separation of information. The Cell pattern can be blended with other patterns to create different visualization techniques, such as treemaps, nested tables, Voronoi diagrams, Venn diagrams, sunburst (multilevel pie) charts, packing circles, Karnaugh maps, house of quality diagrams, and icicle trees. Figure 4.5 shows some Cell-based structures.

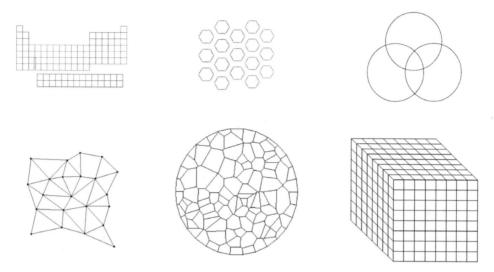

Figure 4.5: Examples of Cell-based structures.

4.6.5   COORDINATE

Designers can use this pattern to map items from an information space onto visualizations and organize them in the representation space with respect to a frame of reference. This pattern can be useful when designers want to represent information items that can be placed and located with respect to external structures. The Coordinate pattern is typically instantiated with axes or scales, which allow for describing the location or position of visualizations with respect to such structures. The lines that form these structures can be straight, curved, or circular, where these can be blended recursively to create complex grid, parallel, spherical, polar, or other organizations. Examples of these structures include one-, two-, or three-dimensional coordinate systems, number lines, Cartesian coordinates, polar coordinates, and spherical coordinates. The Coordinate pattern can organize information in the representation space according to temporal, ordinal, nominal, categorical, conceptual, or spatial value and/or positional frames of reference. This pattern can be blended with other patterns to create many different visualization techniques, such as scatterplots, bubble charts, histograms, ternary graphs, and stacked charts. Figure 4.6 shows some structures that are based on the Coordinate pattern.

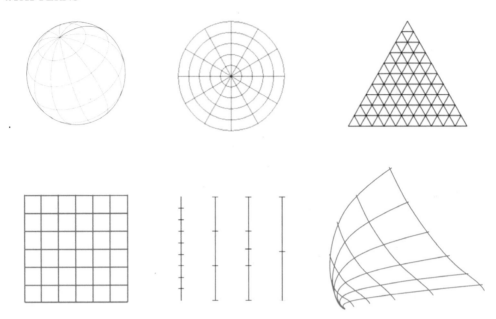

Figure 4.6: Examples of Coordinate-based structures.

4.6.6   CYCLE

When using this pattern, designers map information items onto visualizations in the representation space and organize them in a circular, wheel-like, rotational, spiral, and/or cyclical fashion. This pattern can be useful when a designer wants to represent information items that are periodic and have recurrence, such as can be the case with astronomical, geological, climatic, biological, mathematical, and other similar information spaces. It is important to note here that using a circular structure in a visualization does not necessarily mean that the representation is based on the Cycle pattern. Figure 4.7 shows three structures that are based on the Cycle pattern. This pattern can be blended with others to create different visualization techniques, such as spiral graphs, concentric ring charts, and polar area diagrams.

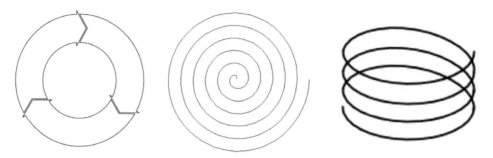

Figure 4.7: Examples of Cycle-based structures.

4.6.7   FUSION

Designers can use this pattern to map items from an information space onto a single visualization in a continuous fashion, such that these items are integrated and fused together in the representation space. This pattern can be useful when a designer wants information items to be represented in such a way that their conjoinment makes them *indistinguishable* from one another. In other words, there are a number of items (e.g., data points) that could be represented individually with different representations, but the designer wants them to be fused together into one integrated structure so they are represented in a continuous rather than discrete fashion. The main difference between the Token and Fusion patterns is as follows. The designer can use Token to map information onto one unitary representation. However, this representation can be composite, made of discrete parts encoding multiple properties or features of data—e.g., a Chernoff face. In using Fusion, however, the designer wants to map multiple items or data points onto one unitary representation that has fused all the items into a singular, continuous whole that has no discernable component parts. Therefore, the Token and Fusion patterns are used for different types of mappings. The Fusion pattern always deals with a multiplicity of items from the information space—never one. A number of common visualization techniques are based on this pattern, including 1-, 2-, and 3-d fitted plots, stacked area charts, streamgraphs, sparkline charts, and some cluster maps. Figure 4.8 shows three structures that are based on the Fusion pattern. These types of visualizations usually represent the integration of a set of data points that form the sub-systems of mathematical plots—visual representations that, although not shown, are typically instances of a blending of the Coordinate and Fusion patterns, where a Coordinate-based structure acts as a substrate in/on which the Fusion-based visualization resides.

Figure 4.8: Examples of Fusion-based structures.

4.6.8   GROUP

When using this pattern, designers map information items onto visualizations and organize them in the representation space by congregating them close to each other. This pattern can be useful when a designer wants to represent proximate features of information items—whether the proximity be semantic, syntactic, spatial, temporal, functional, or otherwise. Figure 4.9 shows three examples of simple structures that are based on the Group pattern.

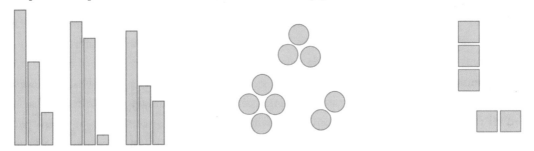

Figure 4.9: Examples of Group-based structures.

4.6.9   HIERARCHY

Designers can use this pattern to map items from an information space onto visualizations and organize them in the representation space in a hierarchical, multi-level, pyramid-like fashion, where higher levels are superior to or contain and encompass lower levels. This pattern can be useful when designers want to represent parent-child, higher-lower, superior-subordinate, whole-part, container-contained, or system-part relationships. Many representational techniques are based on the blending of Hierarchy with other patterns. Common examples are node-link trees, sunburst charts, cone trees, dendograms, and treemaps. Figure 4.10 shows some structures that are based on the Hierarchy pattern.

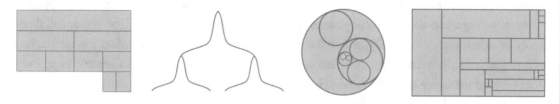

Figure 4.10: Examples of Hierarchy-based structures.

4.6.10 LINK

When using this pattern, designers map information items onto visualizations and organize them in the representation space by connecting them together using paths, routes, lines, or other similar structures. This pattern can be useful when designers want to represent explicit connections, relationships, and/or associations between and among representations, whether temporal, causal, functional, conceptual, epistemological, or otherwise. Many visualization techniques are based on blending Link with other patterns. Common examples are network diagrams, node-link trees, NodeTrix, flow charts, arc diagrams, radial networks, concept maps, and process flow diagrams. Figure 4.11 below shows two simple structures that are based on the Link pattern.

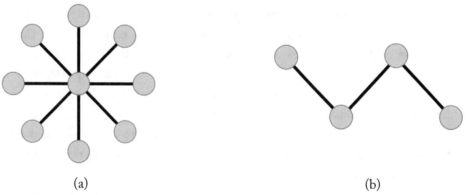

(a)   (b)

Figure 4.11: Examples of Link-based structures.

4.6.11 LIST

Designers can use this pattern to map items from an information space onto visualizations and organize them in the representation space by placing them in a sequential, successive fashion. This pattern can be useful when designers want to represent information items that have a definite order—whether chronological, alphabetical, or otherwise—as well as a clear start and end point. In other words, unlike the Cycle pattern, the start and end points for the instances of this pattern do not join and/or recur in a cyclical fashion. Using this pattern, visual representations are typically placed close together, and can be organized in any direction, whether, left, right, up, down, diagonally, or otherwise. Additionally, unlike the Coordinate pattern, which organizes representations with respect to an external frame of reference, the List pattern organizes them relative to one another with no necessary external frame of reference to give meaning to their positions. Figure 4.12 shows three example structures that are based on the List pattern.

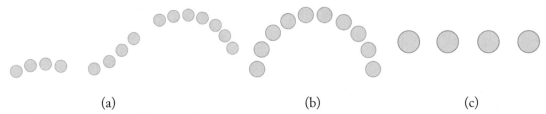

(a)   (b)   (c)

Figure 4.12: Examples of List-based structures.

4.6.12 SPECTRUM

When using this pattern, designers map information items onto visualizations and organize them in a spectral fashion in the representation space. This pattern can be useful when a designer wants to represent variability within a representation or across a number of representations, such as when representing different densities, temperatures, incomes, and ratios. This pattern is often instantiated using multiple saturation or luminance values of a particular hue, especially when the designer wants to represent progressive degrees of variation. This pattern can also be instantiated using multiple hues or textures, usually to represent variation in a more stepped or categorical fashion. This pattern is often blended with others to develop visualization techniques. Examples include choropleths, heat maps, Hertzsprung-Russell diagrams, and spectral maps, plots, and diagrams. Some very basic structures are shown in Figure 4.13 to demonstrate the use of this pattern: they use different saturation values of a single hue to encode variation—some in a continuous and some in a graded or discrete fashion.

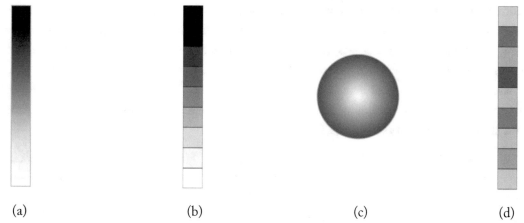

    (a)   (b)   (c)   (d)

Figure 4.13: Examples of Spectrum-based structures.

4.6.13 STACK

Designers can use this pattern to map items from an information space onto visualizations and organize them in the representation space by placing them on top of one another in a piled or stacked fashion. This pattern can be useful when designers want to represent information items that have similar, co-occurrent, or co-existent features. Although not always the case, usually representations are placed on top of one another such that they are touching or are very close together. Common techniques that are based on this pattern include streamgraphs, stacked area charts, and themerivers. Figure 4.14 shows some simple structures that are based on the Stack pattern.

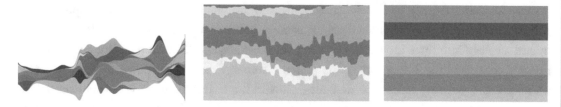

Figure 4.14: Examples of Stack-based structures.

4.6.14 TRACK

When using this pattern, designers map information items onto visualizations and organize them in the representation space in a lane-, stripe-, and/or track-like fashion. This pattern can be useful when designers want to represent information items that exist or occur within routes, paths, or channels, whether conceptually, spatially, temporally, or otherwise. This pattern is also sometimes useful when items have co-occurrence, co-existence, simultaneity, and/or synchrony. Even though information items are placed sequentially along routes, Track does not serve the same mapping function as that of List. The former emphasizes the routes and lanes that happen to give order to items, whereas the latter emphasizes the sequential ordering of items. Lanes can sometimes have no information items on or in them, whereas this is not possible in a List (i.e., it does not make sense to have successive ordering of items when there are no items). Common techniques based on this pattern include swimlane diagrams, genome browsers, and some Vee diagrams. Figure 4.15 shows three structures that are based on the Track pattern.

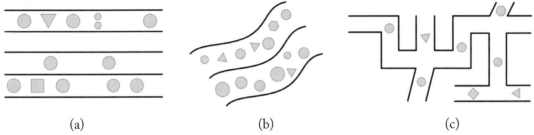

(a)   (b)   (c)

Figure 4.15: Examples of Track-based structures.

4.7   CATEGORIZATION OF PATTERNS

To help guide thinking during design, it can sometimes be helpful to have a categorization of patterns. Because the patterns are abstractions, and are to be used as conceptual tools, they can take on a number of different functions when used in design. As a result, we do not think it is really helpful to provide any rigid categorizations of the patterns here. Designers themselves can categorize the patterns in ways that they find useful. That being said, we propose a categorized grouping of the patterns that we believe is helpful in design situations. In this categorization, we divide the patterns in our language into three groups: (1) primary, (2) substrate, and (3) relational. The first category, *primary*, consists of the Token and Fusion patterns. In many design situations, these two patterns act as primary building blocks for designing visualizations. That is, most visual representations include sub-representations that are based on these two patterns. The second category, *substrate*, consists of the Area, Cell, Coordinate, and Track patterns. These four patterns are often used for designing underlying structures in (on/over) which other representations are placed. Finally, the third category, *relational*, consists of the Branch, Cycle, Group, Hierarchy, Link, List, Spectrum, and Stack patterns. These patterns are often used for creating structures that encode relationships, variations, and/or movements among information items.

Table 4.1 provides a summary of the patterns. Each pattern is characterized and its essential features are given in a compact, gist-like fashion. Readers should consult the previous section for a full description of each pattern.

Category	Pattern	Characterization	Utility
Table 4.1: Characterization of patterns and categories to which they belong			
Primary	Fusion	Map multiple information items onto a single visualization in a continuous fashion, such that the items are integrated and fused together. Most often blended with other patterns, particularly with patterns in the substrate category.	Can be useful when wanting information items to be represented in such a way that their conjoinment makes them indistinguishable from one another.
	Token	Map one or more information items onto a visualization that can be regarded as a unit, whether in atomic form or composite form made of discrete parts. Most used pattern for representing information. Most visualizations are based on a blending of Token with other patterns. Instantiations are often small.	Can be useful when wanting to represent an information item, such as an object, property, value, system, operation, location, direction, feature, function, or process, in an individual, unitized fashion in the representation space.
Substrate	Area	Map information items onto visualizations in such a way that their boundary, shape, region, and/or area are encoded.	Can be useful when wanting to represent items that have two or three spatial dimensions with a definite shape, boundary, or region.
	Cell	Map information items onto visualizations and organize them by segmenting, compartmentalizing, or containing them within cell-like structures.	Can be useful when wanting to create partitions or compartments for a set of items according to their temporal, ordinal, nominal, categorical, conceptual, or spatial properties, and want to represent distributed membership, containment, division, or separation of information.
	Coordinate	Map information items onto visualizations and organize them with respect to a frame of reference. Typically instantiated with axes or scales, which allow for describing the location or position of sub-system visualizations with respect to such structures.	Can be useful when wanting to represent items that can be placed and located with respect to an external structures or frame of reference.
	Track	Map information items onto visualizations and organize them in a lane-, stripe-, and/or track-like fashion.	Can be useful when wanting to represent items that exist or occur within routes, paths, or channels, whether conceptually, spatially, temporally, or otherwise. Also sometimes useful when items have co-occurrence, co-existence, simultaneity, and/or synchrony.

Relational	Branch	Map information items onto visualizations and organize them in the representation space in a branched and/or subdivided fashion.	Can be useful when wanting to represent converging and/or diverging (e.g., temporal, categorical, conceptual, or geographic) features of information items in addition to whether they originate from or terminate in other sources.
	Cycle	Map information items onto visualizations and organize them in a circular, wheel-like, rotational, spiral, and/or cyclical fashion. Using a circular structure in a visualization does not necessarily mean that it is based on the Cycle pattern.	Can be useful when wanting to represent items that are periodic and have recurrence.
	Group	Map information items onto visualizations and organize them by congregating them close to each other.	Can be useful when wanting to represent proximate features of information items—whether semantic, syntactic, spatial, temporal, functional, or otherwise.
	Hierarchy	Map information items onto visualizations and organize them in a hierarchical, multi-level, pyramid-like fashion, where higher levels are superior to or contain and encompass lower level items.	Can be useful when wanting to represent parent-child, higher-lower, superior-subordinate, whole-part, container-contained, or system-part relationships.
	Link	Map information items onto visualizations and organize them by connecting them together using paths, routes, lines, or other similar structures.	Can be useful when wanting to represent explicit connections, relationships, and/or associations between and among items, whether temporal, causal, functional, conceptual, epistemological, or otherwise.
	List	Map information items onto visualizations and organize them by placing in a sequential, successive fashion.	Can be useful when wanting to represent items that have a definite order—whether chronological, alphabetical, or otherwise—as well as a clear start and end point.
	Spectrum	Map information items onto visualizations and organize them in a spectral fashion. Often instantiated using multiple saturation or luminance values of a particular hue, or using multiple hues or textures.	Can be useful when wanting to represent variability within a visualization or across a number of visualizations, such as when representing different densities, temperatures, incomes, and ratios.
	Stack	Map information items onto visualizations and organize them by placing on top of one another in a piled or stacked fashion. Visual representations are often placed on top of one another such that they are touching or are very close together.	Can be useful when wanting to represent items that have similar, co-occurrent, or co-existent features.

4.8 USAGE OF PATTERNS

The 14 patterns presented above are the heart of the visualization design framework in this book. It is important at this point to briefly describe how the patterns are to be used in design, so as to avoid potential confusion. As stated previously, the patterns are abstractions, intended to help designers think about *one-way mapping* from information items to visualizations. The design process that we propose later in this book proceeds from abstract information spaces, down through lower levels of abstraction to the physical instance of a visualization. Because the patterns operate at an abstract level, there is a great deal of flexibility as a designer moves through stages of design—from identifying and blending patterns, to choosing representation and encoding techniques, to deciding on geometric properties. Consequently, designers should be aware that the patterns cannot always be used in a systematic way to analyze existing visualizations. That is, they cannot be used to reverse-engineer visual representations to determine exactly which patterns were blended to create a visualization, as it is not possible to know the intentions of the designer. Thus, while it can be interesting to use the patterns to analyze existing visualizations, and can sometimes shed light on a design process, it can be confusing when using them in a purely analytical fashion for reverse-engineering existing visualizations. There is no way to reverse-engineer the thought process of a designer that led to a given visual representation.

Another important point is that patterns have a polymorphic relationship to visualized instances. An example is given here to demonstrate this point. A designer may generate a visualization that is to serve solely as a representation of the physical boundaries of an information space (e.g., a country). In this case, the visualization can be considered an instance of the Token pattern, particularly if it is small. However, if the same visualization is to serve as a substrate structure for placing other representations on it, then it can be considered an instance of the Area pattern. This simple example highlights the importance of viewing the patterns as conceptual tools for design thinking, rather than as physical patterns that exist in visualizations. Along the same lines of reasoning, it is sometimes possible for a visualization to be arrived at through different blendings of patterns. For instance, if two patterns, P_1 and P_2, are blended to arrive at a visualization, it may be possible to use patterns P_2 and P_3 to come to the same visualization. Again, it is important to keep in mind that the patterns are not simply abstract characterizations of physical entities that share a one-to-one mapping; rather, they are *abstractions that function at the level of design intention*, and are meant to promote systematic design thinking with respect to what it is that the designer wants to convey to his or her audience. As a result, contemplating different patterns in a design situation can trigger similar outcomes, depending on the designer, the information space, the context of design, and so on.

In Section 4.6 we gave examples of a number of different common visualization techniques that are based on a given pattern in order to clarify the function of each pattern. It is very important to note, however, that the labels given to some visualization techniques are often applied to visualizations that are based on different combinations of pattern blendings. For example, two visualizations can be called heatmaps, yet be based on different blendings. Additionally, common techniques often have new twists added to them—for example, a treemap may have node-link or another type of structure embedded within it, and yet still be referred to as a treemap. Such additions change the overall emergent properties of the visualization that are not captured using technique names, yet can be detected and understood when seen through the lens of the presented patterns. Finally, techniques can have different labels, and look quite different from each other in the representation space, yet be based on the same basic pattern blendings and represent the same general information, as will be clearly demonstrated in the next chapter.

The discussion in this section highlights for the designer that approaching and thinking about representations using the abstract patterns in this chapter, rather than their visualization technique names, can help with understanding their fundamental building blocks and representational affordances. A knowledge of these patterns helps designers develop a sense of how to quickly determine the underlying pattern blendings in a concrete visualization. Nonetheless, we downplay this analytical utility of the pattern language in this book because of the polymorphic nature of patterns.

Chapter 5 discusses the next part of the pattern language: how different patterns can be blended together. It helps designers think about how patterns can be blended together, and then instantiate the blendings in any fashion they see fit—whether in a simple, complex, embedded, nested, recursive, or other manner. It will also demonstrate the operational utility of the pattern language through a series of examples.

CHAPTER 5

Blending of Patterns

In the previous chapter, we characterized the 14 abstract patterns that constitute the building blocks or letters of our pattern language—the core element of our visualization design framework. These patterns can guide and support design thinking. In this chapter we describe the second component of the pattern language—i.e., its syntax for blending the patterns. The real power and utility of the pattern language is this: given different information spaces and tasks, designers can use the patterns as tools for thinking about how they can map and organize information in the representation space.

This chapter is organized as follows. Section 5.1 discusses how patterns can be blended—the syntax of the language and its nesting capability. Section 5.2 demonstrates the operational utility of the pattern language by presenting a series of visualizations that exemplify different blendings. Finally, using examples, Section 5.3 shows the utility of the pattern language not only as support for design thinking, but also as a tool for analysis of the structure and composition of visualizations.

5.1 BLENDING OF PATTERNS

Designers can blend different patterns to devise representational structures that have different organizational affordances. They can blend the 14 patterns in numerous ways, using different combinations. Hence, blending the patterns provides designers with basic language-like thinking, making the design of countless representational techniques and visualizations possible. In addition to the metaphor of letters used previously, another analogy is apt here. Consider programming languages. They have constructs such as loops, conditionals, data types, and subroutines. Different languages instantiate these constructs differently. Programmers blend these constructs in innumerable ways to write code for different purposes and applications. Rather than thousands of examples of code templates, programmers can use basic programming constructs as tools for thought to design pseudo code before worrying about implementation details. The visualization design patterns in this book are similar to programming constructs. In a similar fashion, rather than dealing with the thousands of techniques and implementations that are already out there, the patterns and their blendings allow the designer to think about visualizations in terms of a few design constructs that are abstract, yet close enough to concrete instantiations that they can be readily used to build new visualizations.

The next section presents the simple syntax of the language for blending the 14 patterns. The section after that discusses the nesting power of the language.

5.1.1   SYNTAX

To represent the blending of the different patterns, our pattern language uses a simple syntax. The syntax has three elements: (1) 14 codes (e.g., TK and CR) to denote the different patterns (see table below for a list of the codes); (2) the symbol "•" to denote a blending, where blended patterns appear together in square brackets "[]"; and (3) the symbol "∈" to denote that a visualization or representational structure "is derived from," "is based on," "instantiates," or "is an instance of" a blending. Hence, the expression $V_i \in [P1 \cdot P2 \cdot P3]$ signifies that the visualization, V_i, is an instance of or is derived from the blending of the three patterns P_1, P2, and P_3, where *the order of the patterns is of no consequence.* This is the case because the patterns and their blendings are abstractions, and we do not want to reify the abstract construct of patterns by blending them in an ordinal fashion—that is, giving precedence to one pattern over another. Since patterns are abstractions, we can also say that a visualization technique is "based" on a blending of patterns. Representations can be derived from and be instances of a single pattern or the blending of several patterns (theoretically, up to a blending of all 14 patterns). Therefore, to make our sentences shorter, sometimes in referring to a visualization or structure that is based on or is an instance of a blending, we merely use the term "$[P_i \cdot P_j \cdot P_k]$-based visualization," or state that the structure is "$[P_i \cdot P_j \cdot P_k]$-based," or state that a visualization "instantiates $[P_i \cdot P_j \cdot P_k]$."

Pattern	Code	Pattern	Code	Pattern	Code	Pattern	Code
Area	AR	Cycle	CC	Link	LK	Token	TK
Branch	BR	Fusion	FS	List	LS	Track	TR
Cell	CL	Group	GR	Spectrum	SP		
Coordinate	CR	Hierarchy	HR	Stack	ST		

5.1.2   SELF-BLENDING AND NESTING

As we stated before, apart from atomic visual marks, other more complex visualizations or structures are obtained by composing simpler structures. Sometimes a pattern can be self-blended to create composite structures. When representing self-blendings, if we denote the first structure derived from pattern P, as V_1 and the subsequent ones as V_2, V_3, … Vi, the final composed representation, V, is simply an instance of P, so its expression would be written as $V \in P$ and not $V \in [P \cdot P \cdot P \cdot …]$. An application of self-blending is in nesting of self-similar structures. A simple self-similar nesting, for example, involves the Cell pattern. A simple table-like structure ($\in CL$) can be nested in another structure of the same type. It is possible to imagine a table whose cells contain other tables, whose cells contain other tables, whose cells contain other tables, and so on. This same type of example can be imagined for other patterns such as Area, Coordinate, List, and Track.

In addition to self-similar nesting, structures can overlap one another, can be placed adjacent to each other, can be layered on top of each other, and can be placed close to or distant from one another in the representation space. These juxtapositions can be repeated at many levels of nesting complexity. A simple example of this is a multidimensional glyph, such as a Chernoff face. This type of glyph results from combining structural instances of the Token pattern (e.g., lines, dots, and geometric shapes) together to create more detailed instances of the same pattern. Designers can nest instances of all kinds of blended patterns in other structures. For instance, in design thinking, a set of icons, instances of Token, can be placed on a polar coordinate structure ($\in$ CR), that is placed within the cell of a partitioned structure ($\in$ CL), that is placed as an element of a sequential structure ($\in$ LS), that is placed within a map-like structure ($\in$ AR), and so on. What we call such structures or how we perform post-design analysis of final structures that result from these blendings is of no consequence. What matters is how the patterns help the designer to create such elaborate visual structures with which users can interact to carry out complex cognitive activities. This nesting feature of the pattern language of our framework is a direct result of thinking about the information and representation spaces in terms of the properties of systems (i.e., systems, sub-systems, etc.), as discussed in Chapter 3. These nested structures can play a powerful role in the context of interactive visualizations representing complex information spaces that contain many sub-space levels. Initially, only the high-level space may be encoded. Subsequently, upon interacting with the visualizations, deeper levels of these sub-spaces can be encoded and unfolded progressively.

5.2   OPERATION OF PATTERN LANGUAGE

In this section we provide a series of examples[8] to demonstrate the operational utility of the pattern language—that is, how the patterns can be blended to create different visualizations and representation techniques. Three important points should be kept in mind. Firstly, and most importantly, since we cannot go through a design process for each example from scratch, we have chosen concrete, finished visualizations (including their detailed visual marks and variables) as examples. As a result, we have adopted an analytical approach in discussing many of the examples, even though we have stated previously that the main application of the framework is not to reverse-engineer the design process of visualizations to determine their underlying patterns. This is because *the designers' intentions and thinking behind why a visualization was designed in a certain way are inaccessible to us.* Indeed, many visualizations may not readily lend themselves to post-hoc analysis. Yet we knowingly violate the prime intention of the framework, which is to guide design thinking, in order to demonstrate that innumerable techniques and visualizations can be based on pattern blendings. Secondly,

---

[8]   We do not necessarily endorse the examples in this chapter as excellent designs. A number of them could probably be improved in various ways. The main point of including them here is to demonstrate possibilities for pattern blendings. Lower-level concerns related to visual variables, layouts, and so on are not addressed in this chapter. See Chapter 7 for more on this.

as there are countless numbers of visualizations that can be created, we do not give examples of all possible blendings—that would make this book very lengthy and unmanageable. Rather, we provide examples that demonstrate what we mean by pattern blending; how simple, yet effective, pattern blending is as a conceptual design tool; and the wide range of and diversity of techniques that can be generated using different pattern blendings. We focus on examples that illustrate the utility and flexibility of our visualization pattern language. Many of the examples below may seem simple because they are instances of the blending of only a few patterns. However, they exemplify the generative nature of the language and how more complex visualizations can be created using the suggested 14 patterns. Thirdly, and finally, when studying the following examples, the reader must adopt a designer's perspective. That is, rather than focusing on the semantics of the representations, the reader should focus on the syntactic nature of the blendings and their combinatorial flexibility, lending to possibilities for design thinking. In the following sections, we present the patterns in the order of the categories to which they belong. Our approach is motivated by the desire to make it easier for readers to understand the role that the patterns play in the blendings. Before proceeding, the reader is encouraged to review the pattern codes (e.g., AR, CR, TK) that are used in the syntax of the language, as they will be used extensively in the design and analysis of visualizations in the following sections.

5.2.1   PRIMARY PATTERNS

Blendings of Token (TK)

As a primary pattern, Token has usage in the majority of blendings. Here we will provide some simple examples of a few of its blendings with other patterns. These can help the reader develop a better understanding of blendings in the later sections of this chapter.

First, let us see how we can self-blend the Token pattern. Figure 5.1a shows a simple example of self-blending. The representation is composed of two sub-representations: (1) the outline of a human body ($\in$ TK); and (2) the human skeletal system ($\in$ TK) that is placed on top of the outline (i.e., the first sub-visualization). The skeletal sub-representation, itself, is a composite representation. It is made of a set of sub-representations that are themselves made of smaller visual marks ($\in$ TK) that have been composed together. The overall visualization is an instance of the self-blending of Token at several levels. However, we simply refer to it as Token-based.

Figure 5.1b and 5.1c show two [TK•GR]-based visualizations, where 5.1b shows a clustering ($\in$ GR) of a set of glyphs ($\in$ TK; self-blended), and 5.1c is a clustering ($\in$ GR) of a set of images of planets ($\in$ TK).

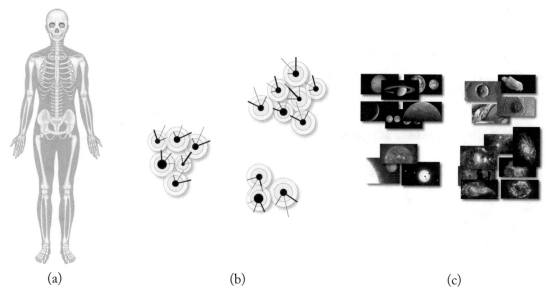

(a)   (b)   (c)

Figure 5.1: Examples of TK- and [TK•GR]-based visualizations. Image in (a) courtesy of Matthew Cole/Shutterstock.com.

Figure 5.2 shows three [TK•LS]-based visual representations. The first visualization shows, from left to right ($\in$ LS), the first five cuneiform letters ($\in$ TK) of Sumerian. As can be seen, the cuneiform letters are all composed of smaller visual marks. The second visualization, from left to right, depicts the successive stages ($\in$ LS) of the gradual truncation of the vertices and edges of a cube to demonstrate how a rhombitruncated cuboctahedron ($\in$ TK) can be obtained from the cube (Sedig et al., 2003). The visualization technique used in the second representation is referred to as the small multiples. The third example shows an arithmetic expression, where all the mathematical symbols are Token-based and are sequentially organized ($\in$ LS). Even though self-blending is present in all three examples here, we will not repeatedly mention it anymore, as the concept of a self-blended pattern should be clear by now.

$$\sum_{i=1}^{5}(i-1) = 0+1+2+3+4 = 10$$

Figure 5.2: Examples of [TK•LS]-based visualizations.

Figure 5.3a shows a [TK•CL]-based visualization, composed of a 2D hexagonal hive-like structure (∈ CL) and a set of numbers (∈ TK) that are distributed and placed in its cells. Figure 5.3b shows a [TK•AR]-based visualization, composed of an analogical (geographic) map (∈ AR) showing the shape and boundary of France and discrete names (∈ TK) of places on the map. Figure 5.3c shows a [TK•CR]-based design, composed of a Cartesian coordinate structure (∈ CR), along with individualized visual marks (∈ TK)—i.e., plot points that encode other items in an information space. This representation technique is called a scatterplot.

(a)   (b)   (c)

Figure 5.3: Examples of [TK•CL]-, [TK•AR]-, and [TK•CR]-based visualizations.

Figure 5.4a shows a [TK•BR]-based visualization, usually referred to as a Sankey diagram. This diagram can be interpreted in two ways: an information space that gets broken down (diverge) into sub-spaces, or a set of spaces that converge and form a larger space. This diverging or converging is usually encoded using a set of weighted lines (∈ BR) along with nodes encoding the unitized sub-spaces (∈ TK).

Let us consider an information space whose items are geometric objects (e.g., Archimedean and Platonic solids). Our goal is to design a visualization that shows how a set of Archimedean shapes can be obtained from a base Platonic solid (e.g., a cube) by truncating its vertices and edges (Sedig et al., 2003). Here, from description of the problem, we can readily see that we can use Token and Link. Furthermore, since our data have directional relations, we can use arrows or pointers. Figure 5.4b shows a possible [TK•LK]-based design for this problem. This visualization is composed of shapes that analogically encode the items (solids) in the information space (∈ TK), lines that encode connections between adjacent solids (∈ LK), and arrows (∈ TK) that encode the directions of how solids can be obtained from each other.

Finally, Figure 5.4c shows a [TK•SP]-based visualization, composed of an analogical shape encoding a hand—a singular, individualized unit of information—(∈ TK), combined with spectral colors (∈ SP) encoding variation in temperature in different parts of the hand. The visualization represents the thermal imaging of a hand—i.e., it is a heatmap.

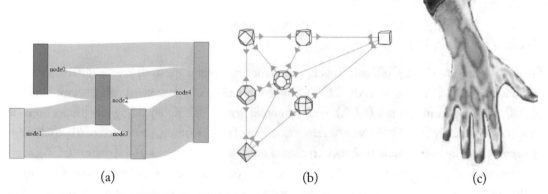

(a)   (b)   (c)

Figure 5.4: Examples of [TK•BR]-, [TK•LK]-, and [TK•SP]-based visualizations.

Blendings of Fusion (FS)

The Fusion pattern maps a multiplicity of information items onto a visualization in a continuous way that makes the items indistinguishable from one another. Figure 5.5a shows a mathematical knot composed of two visual structures whose algorithmically generated information items are fused together in an indistinguishable continuous whole (∈ FS). The knot is displayed against two frames of reference (∈ CR), one of which uses a Token-based arrow to show the orientation of the knot. Figure 5.5b shows a [FS•SP]-based visualization. This representation is called a carpet plot. The dataset is a 2D array whose data points are fused together into a continuous plot (∈ FS), and variations in peaks and troughs of the plot are encoded in a spectral fashion (∈ SP). In this example, the frame of reference is implicit and not explicitly encoded.

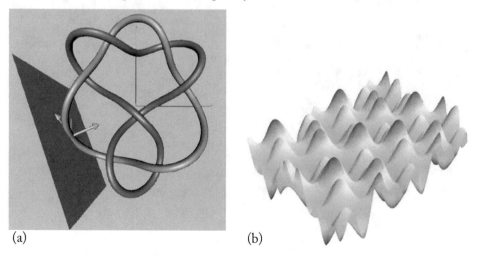

(a)   (b)

Figure 5.5: Examples of Fusion-based visualizations. (a) Used with permission and adapted from KnotPlot (Scharein, 2015). (b) Courtesy of Mayavi.

5.2.2   SUBSTRATE PATTERNS

Blendings of Area (AR)

Given the patterns Area, Cell, and Token, a designer can blend these to create different visualizations. Figure 5.6a shows an [AR•CL•TK]-based visualization, composed of a geographic map
(∈ AR), partitioned into states (∈ CL), with symbolic glyphs (∈ TK) encoding the proportions of
people's weights. Figure 5.6b shows a different example of this blending, composed of a bounded region encoding the overall brain (∈ AR), partitioned into different sections of the cortex (∈ CL), such
as premotor and primary motor, that are all numbered (∈ TK). This visualization is an illustration
of the Bromman's areas of the neocortex. The designer may want another variation of this blending
by adding spectrum. Figure 5.6c shows an [AR•CL•TK•SP]-based visualization, composed of a
geographic map of the U.S. (∈ AR), its states (∈ CL), their counties (∈ CL), color encoding value/
rate (∈ TK), displayed over the country in a spectral fashion (∈ SP). This visualization is a map of
the kidney, renal, and ureter cancer mortality rates by county. As we have seen, this visualization
technique is commonly called a heatmap. A very similar visualization, with a variation in blending,
[AR•CL•SP], is seen in Figure 5.6d, representing the spatial distribution of temperature over the
U.S. Unlike the previous one, in this representation, we do not need the individual values attached
to each cell in the partition. This visualization is composed of a geographic map (∈ AR), which is
divided into states (∈ CL), and uses a spectral coloring (∈ SP) to show variation and distribution
of the data, overlaid on top of the entire map. This visualization technique is sometimes referred to
as a choropleth map.

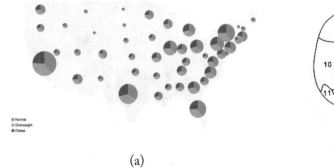

(a)

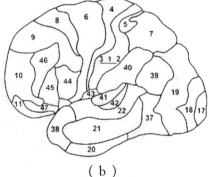

(b)

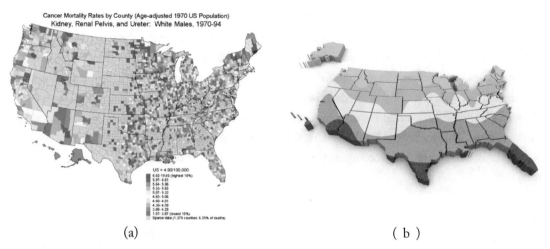

Figure 5.6: Examples of visualizations based on blendings of the Area pattern. (a) From Heer et al., (2010). (b) Courtesy of Prof. Mark Dubin, University of Colorado-Boulder. (c) From Devesa et al. (1999). (d) Courtesy of iQoncept/Shutterstock.com.

Blendings of Cell (CL)

As a substrate pattern, Cell can be used in the design of many visualizations. In this section, we look at a few blendings of this pattern. Figure 5.7a shows a [CL•TK]-based visualization, a nested diagram composed of a partitioned structure (∈ CL) with individual set members (∈ TK) placed in its compartments. As we stated previously, it is important to note that we can self-blend Cell. As such, a Cell-based structure can be self-nested (as in the case with nested tables). Nonetheless, the final structure is still an instance of Cell. Hence, a nested Venn diagram, no matter how complex and nested, is still [CL•TK]-based. A designer can choose to add the Spectrum pattern to this basic blending of Cell and Token to get [CL•TK•SP]. Figure 5.7b shows a [CL•TK•SP]-based hexagonal heatmap, composed of a 2D hive-like structure (∈ CL), with colors in the cells encoding values of items (∈ TK). The variation of these values over the entire space is then encoded in a spectral fashion (∈ SP).

Instead of Spectrum, as in the previous blending, a designer may choose to add the Hierarchy pattern to [CL•TK], resulting in [CL•HR•TK]. Variations of this pattern allow the creation of some popular representational techniques, two of which are presented here. The first technique is that of a mosaic plot, often a visualized rendition of nested tables dealing with categorical data that are hierarchical (Figure 5.7c). This type of mosaic plot is [CL•HR•TK]-based, where the size of each cell encodes the statistical value of an information item and its color encodes its categorical type (∈ TK). We can obtain a more elaborate visualization technique from this blending, the tree-map technique, where a designer may also need to show variation and fluctuation of information items. Figure 5.7d shows a [CL•HR•TK•SP]-based treemap, composed of nested partitioning (∈ CL), parent-child, whole-part organization of the data (∈ HR), individualized values and labels for cells (∈ TK), and spectral encoding to show distributed variation of information over the whole expanse of the visualization (∈ SP). As seen, a designer can blend the patterns of cellular heatmaps (i.e., [CL•TK•SP]) with the Hierarchy pattern to obtain treemaps. Finally, Figure 5.7e shows a [CL•HR•TK]-based visualization—i.e., a Voronoi diagram for the average food spending of Americans. This technique is a hierarchical partitioned structure (∈ [CL•HR]) with food names (∈ TK) in it, but rather than rectangular structures, as in the case of treemaps, it often uses polygonal cells.

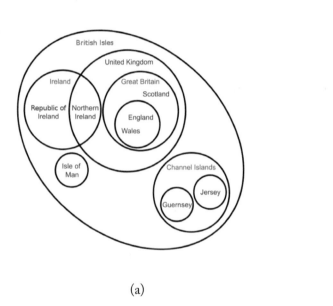

(a)

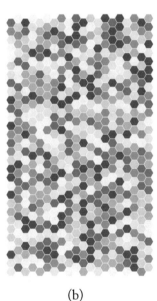

(b)

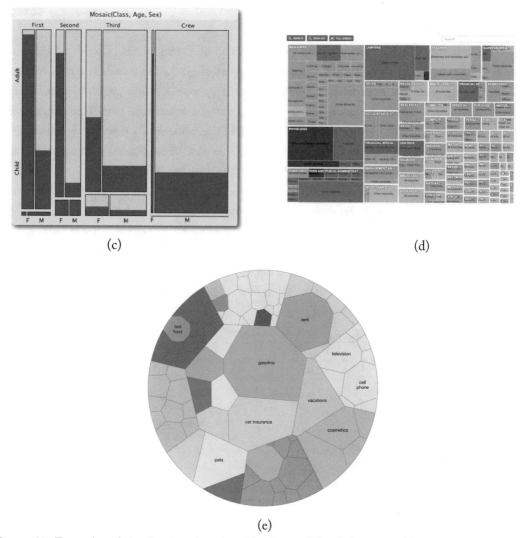

Figure 5.7: Examples of visualizations based on blendings of the Cell pattern. (c) Courtesy of Martin Theus and Mondrian (www.theusRus.de/Mondrian).

Blendings of Coordinate (CR)

As a substrate pattern, similar to Cell and Area, Coordinate can be used in the design of many visualizations. Let us see some visual representations that can be based on blendings of Coordinate with other patterns. We have already seen a representation based on [CR•TK] (Figure 5.3c). Figure 5.8a shows yet another representation that can be obtained from this blending—an abstract visualization that can be used in different contexts (e.g., computer games). It is composed of a grid (∈ CR) with exact grid locations where objects (∈ TK) are placed, and arrows (∈ TK) that encode the direction of possible movement of objects.

We can vary this blending and use Fusion instead of Token to get [CR•FS]. Figure 5.8b shows a [CR•FS]-based visualization, composed of a set of graduated axes laid out in a circular fashion (∈ CR), plus plot lines that encode coalescence of a set of data points into continuous wholes (∈ FS). This representation is usually called a circular coordinate plot. This structure can inspire the creation of star plot glyphs (∈ TK) placed in the context of a frame of reference—i.e., a Cartesian coordinate system (∈ CR) (Figure 5.8c). This new visualization, a scatterplot, is also [CR•TK]-based, and is more complex than the one we saw before (Figure 5.3c).

Now, consider a relatively large multidimensional dataset whose information items have exact locations (values) in a space. A designer may want to encode not only their locations, but also show their trends of variation. We can vary [CR•FS] by adding the Spectrum pattern to it. The visualization in Figure 5.8d can be generated from [CR•FS•SP], and is composed of a set of scalar parallel planes (∈ CR), plot lines (coalescing of data points) each encoding a single observation (∈ FS), and spectral presentation of variation in the data (∈ SP). This visualization technique is called the parallel planes coordinate plot.

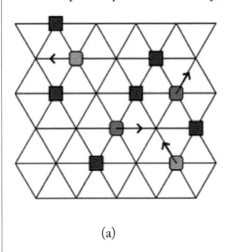

(a)

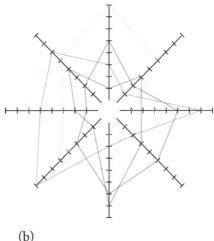

(b)

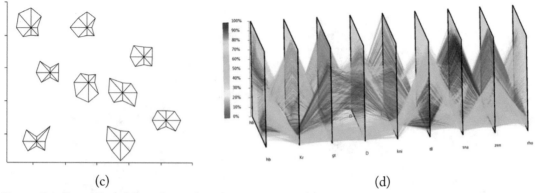

(c) (d)

Figure 5.8: Examples of Coordinate-based visualizations. (d) Courtesy of © 2015 The regents of the University of California, through the Lawrence Berkeley National Laboratory.

Rather than Spectrum, a designer may want to blend [CR•FS] with Token and Group. A simple, yet very interesting [CR•FS•TK•GR]-based visualization is seen in Figure 5.9. The designer has put an area graph ($\in$ [CR•FS]) at the center. Bible chapters (e.g., Mark 16) and the titles of these chapters (e.g., The Resurrection) ($\in$ TK) are used as indices (labels) for the graduated positions on the circular coordinate structure. Two major periods of history are grouped into separate clusters (e.g., Christian and Jewish) ($\in$ GR).

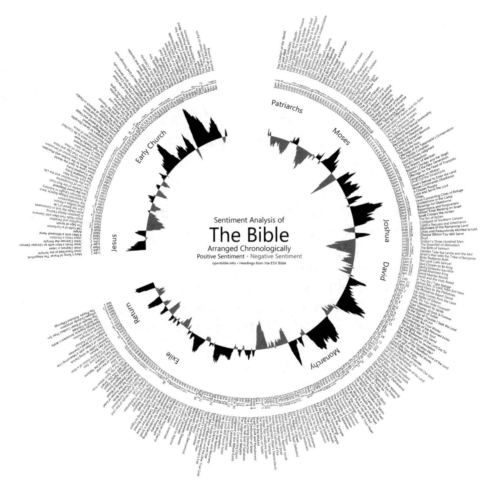

Figure 5.9: Example of a [CR•FS•TK•GR]-based visualization. Courtesy of openbible.info.

We can use [CR•FS•TK•GR] in the previous example to obtain a completely different-looking visualization. The visualization in Figure 5.10 shows space launches by country and purpose, annually. In this case, the designer has used a vertical axis ($\in$ CR) to distribute and place the launches ($\in$ TK) at a graduated range of locations according to year (in this case, 1957 to 2011), with annual payloads for each year encoded as coalesced plot lines ($\in$ FS). The designer has chosen to place the USSR/Russia and the U.S. close to each other and far from the other countries ($\in$ GR), likely due to their high number of launches relative to other countries. In addition, although the individual launches for each country and year could have been organized in a random manner, the designer has chosen to use the Group pattern and organize visualizations by placing military launches together, commercial launches together, and so on ($\in$ GR), which encodes the functional proximity of the launches.

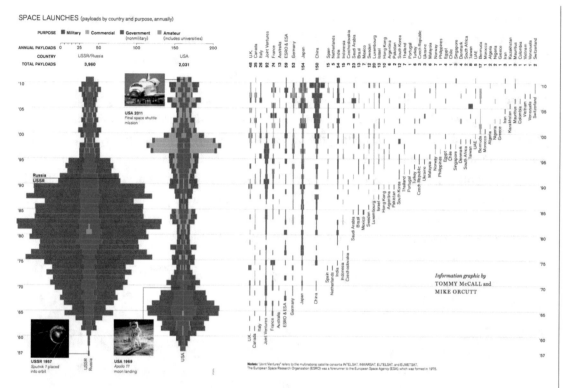

Figure 5.10: Example of a [CR•TK•FS•GR]-based visualization. © 2016. Technology Review. 121014:0216BN.

Sometimes, a designer may want to blend Coordinate and Cell patterns to leverage the structural affordances that both offer. These two can then be blended with Token to create simple and effective representations. A simple example of a [CR•CL•TK]-based visualization is a chess game board (Figure 5.11a), composed of an 8x8 matrix, whose cells have exact indexed positions in the structure (∈ [CL•CR]), where chess pieces (∈ TK) can be placed. We can add the Group pattern to [CR•CL•TK] to get yet another well-known visual representation, the periodic table of elements (Figure 5.11b). The [CR•CL]-based, table-like structure in the representation is used as a frame of reference for locating (i.e., using tuples such as hydrogen-1,1; lithium-1,2; and phosphorous-15,3) chemical elements (∈ TK) in the partitioned cells. Elements with similar properties are then put close together (∈ GR). Hence, the resulting visualization is [CR•CL•TK•GR]-based. Finally, consider an example that is based on [CR•CL•TK], but with an added twist to the nature of the blending. A designer may want to create an indexed table (∈ [CR•CL]) in which to place other representations that can be compared. However, the representations placed in the cells of the table can themselves be based on other blendings. Figure 5.11c shows an example visualization that uses a 2D Cell-based table (∈ CL) that has a small map placed in each of its cells. The table is indexed to give its items exact tuple-based (e.g., <Temp, Apr>) locations (∈ CR). The visualizations in the cells display repeated maps of North America (∈ AR) that use colors and saturation to show variations in three information items (i.e., Net SW, Net Radiation, and Temperature) during each month of the year. Hence, the contained visualizations are [AR•SP]-based, and the structure in which they are embedded is [CR•CL]-based. This representation illustrates how designers can use the visualization pattern language to think about nested and embedded structures using the 14 basic patterns at their disposal, demonstrating the power and flexibility of the language. Furthermore, it is important to keep in mind that we can easily create alternative designs by experimenting with and blending other patterns. For instance, to achieve the same visualization, rather than [CR•CL] as the substrate blending, we can use [TR•LS]. As such, rather than placing the maps in the cells of the table, we can have three lanes on which sequential lists of the [AR•SP]-based visualizations are placed, giving us a representation which is [TR•LS•AR•SP]-based.

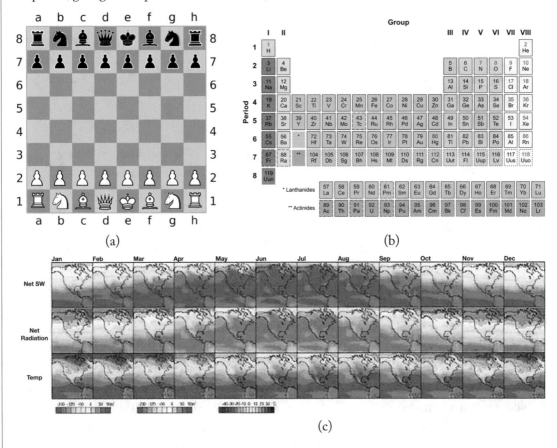

Figure 5.11: Examples of [CR•CL]-based visualizations. (a) Courtesy of Beao, commons.wikimedia. org/wiki/File:Chess_board_blank.svg. (b) Courtesy of Armtuk, commons.wikimedia.org/wiki/File:Periodic_Table_Armtuk3.svg. (c) Courtesy of quince.infragistics.com.

Blendings of Track (TR)

Being a substrate pattern, Track is very useful for organizing information items. We present six-Track-based visual representations that demonstrate this pattern's diverse applications and blendings. Figure 5.12a shows a visualization of a simple children's game. It is composed of four parallel lanes ($\in$ TR) along which a set of different geometric shapes ($\in$ TK) that move down the lanes. As the shapes reach the end of the lane, they are sorted into groups ($\in$ GR) of shapes of the same type in the tray. The visualization also contains lane numbers ($\in$ TK) and an arrow ($\in$ TK) encoding the direction of the movement of the shapes.

Figure 5.12b shows the subway map. This map is [TR•LS•TK]-based. In this visualization, the designer has placed the stations ($\in$ TK) inside different track-like stripes ($\in$ TR). This way the design emphasizes the existence of multiple tracks that run in parallel, sometimes with multiple instances of the same station on the parallel tracks. Stations on the different stripes form several successive ordered lists ($\in$ LS), with start and end stations that form different routes. Hence, this type of map is [TR-LS-TK]-based.

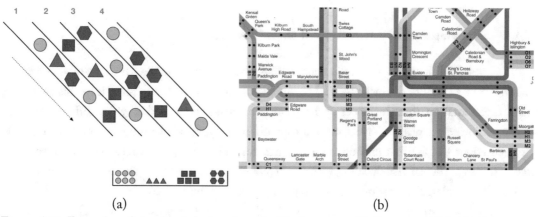

(a)   (b)

Figure 5.12: Examples of visualizations based on the Track pattern. (b) Courtesy of Max Roberts, www.tubemapcentral.com.

Figure 5.13a shows a visualization that is used for planning projects—deliverables, dependencies, and milestones over time. This style of visualization is commonly referred to as a swimlane chart technique. In this representation, there are two main horizontal lanes, one for Project 1 and one for Project 2 ($\in$ TR). Project 2 is then broken down into a number of sub-lanes for different components of the project ($\in$ [TR•HR]). Deliverables ($\in$ TK) are placed within the lanes. Dependencies between deliverables are encoded using links ($\in$ LK) with arrows encoding their direction ($\in$ TK). The overall visualization is thus [TR•HR•LK•TK]-based.

Figure 5.13b shows a set of financial markets and their performance over time. In this visualization, sub-visualizations are placed into four lanes ($\in$ TR), each one for a different market. Within each lane is a sub-visualization—i.e., a horizon graph/chart, where the market's fluctuation over a one-year period is encoded using fused, continuous plots of the data that are spectral ($\in$ [FS•SP]) along a temporal scale ($\in$ CR). The temporal scale is labeled ($\in$ TK). Hence, the overall visualization is [TR•CR•FS•SP•TK]-based.

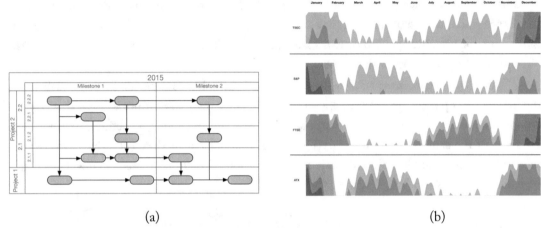

(a)   (b)

Figure 5.13: Examples of visualizations based on the Track pattern.

Figure 5.14 shows a set of audio features that make up an audio file. This type of visualization makes use of a number of tracks ($\in$ TR) in which sub-visualizations are placed. In this particular visual representation, three instances of the Track pattern are being used in which different features of the audio file are placed. Visualizations with different types of blendings—typically plot-style representations ($\in$ CR) that encode audio frequencies ($\in$ FS)—can be placed within the separate tracks. In this particular visualization, we see two plot-like representations stacked within each track ($\in$ ST). The whole collection of tracks is typically given a temporal frame of reference along the horizontal dimension ($\in$ CR). The overall visualization is thus [TR•CR•FS•ST•TK]-based.

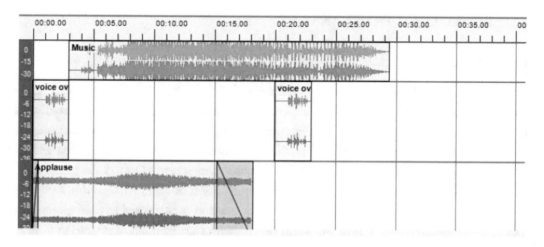

Figure 5.14: Example of a [TR•CR•FS•ST•TK]-based visualization.

Finally, Figure 5.15 shows two variations of a Track-based visualization which shows a map of a region of a city. The visualization shows the record of a person's travel as it is captured by a GPS-enabled step counter. In one of the representations (L) the person's steps are shown discretely and individually (∈ TK), whereas in the other one (R) the step data points are fused together and shown as a continuous line (∈ FS). Hence, the left one is [AR•TR•TK]-based and the right one is [AR•TR•FS•TK]-based.

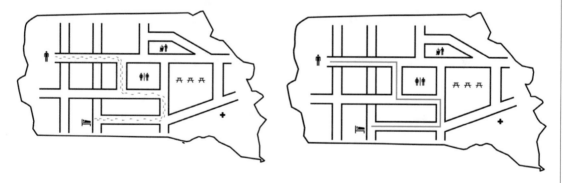

Figure 5.15: Two variations of a Track-based visualization. The left one is [AR•TR•TK]-based and the right is [AR•TR•FS•TK]-based.

5.2.3   RELATIONAL PATTERNS

Blendings of Branch (BR)

We have already seen a [BR•TK]-based visualization (Figure 5.4a). [BR•TK] can be combined with other patterns as well. For instance, Figure 5.16a shows a [TK•BR•CR•GR]-based visualization. The designer has encoded the convergence (merging) of banking groups, where we can see a temporal coordinate structure ($\in$ CR) depicting when different banks ($\in$ TK) merged and four Sankey diagrams ($\in$ BR) that show four different groups of mergers ($\in$ GR). A designer can also blend BR•TK with the Area pattern to create the visualization in Figure 5.16b. This representation shows the migration of anatomically modern humans and is composed of a geographic map ($\in$ AR) divided into different areas of the world ($\in$ AR), with starting point of migration from Africa branching to different parts ($\in$ BR), along with labels/icons ($\in$ TK) as well as arrows showing direction of movement ($\in$ TK). The visualizations showing migration movement are layered on top of the map. Hence, even without knowing the intentions of the designer, it can be seen that, in its basic use of patterns, the visualization is [AR•BR•TK]-based.

Now, consider the use of Branch in representing a materials flow situation in the context of a residential project. There is a wide residential area in which different spaces (e.g., utilities, green space, wetland, and dwelling) should be placed. These spaces house processing units within them (e.g., shower, laundry, condensation, and cooking) that have convergent and divergent in- and outflow of different types of materials (e.g., electricity, water, food, waste, and so on). The designer may decide to use the following patterns for the different parts of the project: Area for the total residential area, Area for spaces, Token for processing units, and Branch for convergent and divergent materials flow. Hence, the resulting visualization shown in Figure 5.17 is [AR•TK•BR]-based.

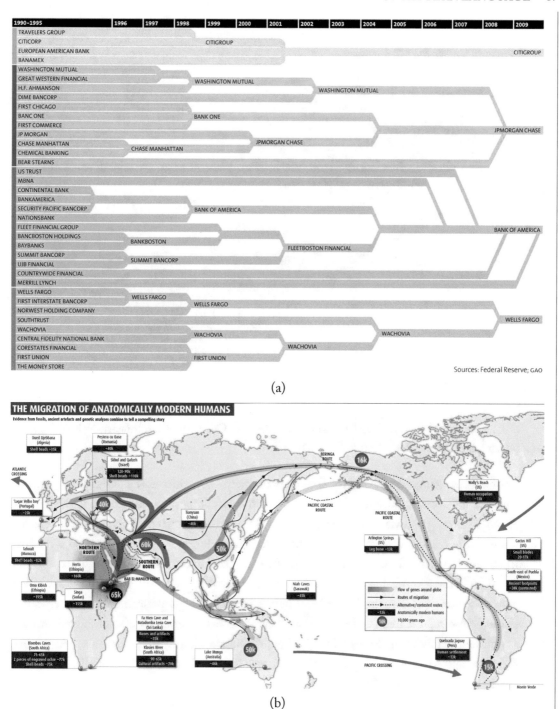

(a)

(b)

Figure 5.16: Examples of visualizations based on blendings of the Branch pattern. (a) Courtesy of Karen Minot. (b) Courtesy of New Scientist.

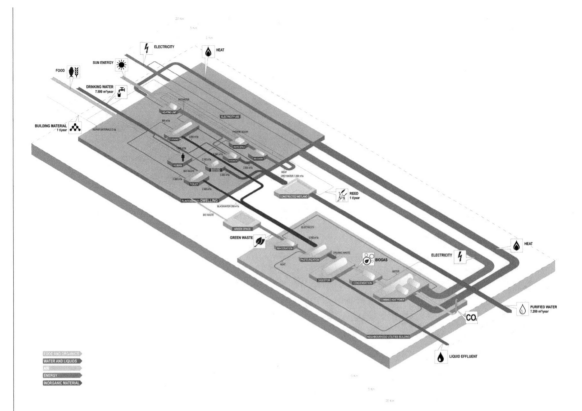

Figure 5.17: Example of a [AR•TK•BR]-based visualization. Courtesy of Jan Jongert, Nels Nelson, Anna Brambilla, Cyclifiers.org, Rotterdam, 2010.

For our last example involving the Branch pattern, let us consider how a designer may encode the evolution of the vertebrata animal kingdom over time, from the Cambrian period to the present. The designer has created a visualization (Figure 5.18) using the Branch pattern in order to encode the common origin and divergent nature of the vertebrata evolution ($\in$ BR). In doing so, a number of information items are fused together to encode each class in a continuous, conjoined manner ($\in$ FS). Additionally, a vertical axis is used to give an external frame of reference to the evolution with respect to time ($\in$ CR). The hierarchical nature of the geologic periods and their respective parent eras is also represented ($\in$ HR). The small symbols at the terminus of each branch section ($\in$ TK) encode the following classes (L to R): Agnatha, Chondrichthyes, Osteichthyes, Amphibia, Reptilia, Aves, and Mammalia. The two extinct classes are Placodermi and Acanthodii. Hence, the visualization is [BR•FS•CR•HR•TK]-based.

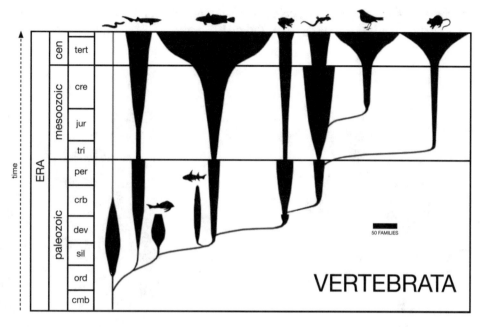

Figure 5.18: Example of a [BR•FS•CR•HR•TK]-based visualization. Adapted from Figure 2 in Benton (1998).

Blendings of Cycle (CC)

Figure 5.19a shows our first example, a [CC•TK]-based visualization that depicts the phases of the moon. It is composed of symbols of the moon (∈ TK) that are placed in a circular fashion to encode the recurring, cyclical phases of the moon (∈ CC). Now, let us examine a self-blended Cycle example. Figure 5.19b shows such a visualization composed of three interlocking Mayan calendars (∈ CC) with symbols for recurring units (∈ TK) in addition to their relative directions (∈ TK). In addition, one of the calendars is nested within another calendar. One of the typical blendings of Cycle is with Coordinate. A simple example of a [CC•CR•TK]-based visual representation is the circular analog clock, shown in Figure 5.19c. We can create more elaborate [CC•CR•TK]-based visualizations by a simple change of the CC-based structure from a circle to a spiral. Figure 5.19d shows such an example. This visualization represents traffic events during a week for a period of a few months. It is composed of a spiral to encode the recurrence of days of the week (∈ CC), graduated rings, moving from inner rings toward outer rings, in a parallel fashion to encode time intervals from month to month (∈ CR), graduated arc segments between the days of the week to encode hours during a day (∈ CR), small individual slices at different hours of the day (∈ TK) to

encode traffic events (e.g., congestion, collision, road work, etc.), all shown in a spectral fashion
($\in$ SP). Hence, this visualization is [CC·CR·TK·SP]-based.

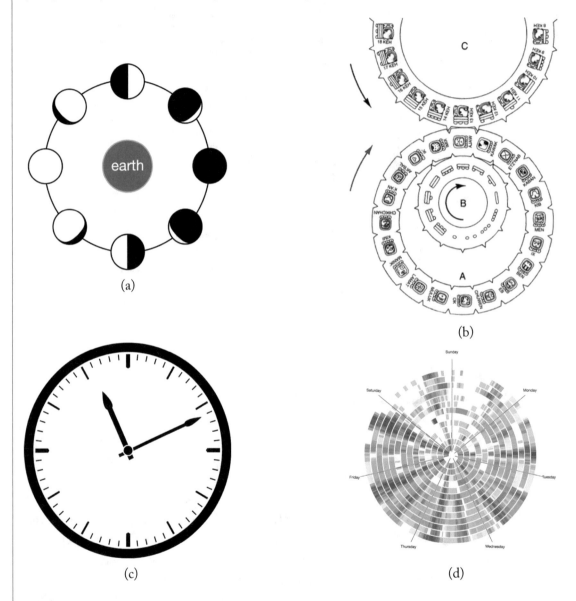

(a)

(b)

(c)

(d)

Figure 5.19: Examples of visualizations based on the Cycle pattern. (b) Courtesy of theedgeoffor-ever2012.com. (d) Courtesy of Maryland CATT Laboratory, www.cattlab.umd.edu.

Finally, a designer can blend Cycle with other patterns to get different effects. Figure 5.20
shows a [CC·CL·GR·TK]-based visualization representing the foods that people should eat each

season, where the circular structure encodes the periodicity ($\in$ CC) of the seasons, and the parti-tioned structure ($\in$ CL) distributes where different food symbols ($\in$ TK) are placed. Seasonal foods are clustered together ($\in$ GR) to suggest when different foods should be eaten.

Figure 5.20: Example of a [CC·CL·GR·TK]-based visualization. Courtesy of Eat Seasonably Calendar © Behaviour Change Ltd. 2015.

Blendings of Link (LK)

Consider an information space whose items have to be organized sequentially. Furthermore, some of these items have to refer to one another—an example being a book that has a set of chapters. In this case, a designer can use [LK·TK·LS]. Figure 5.21a shows an example of this. The small num-bered circles encode the chapters ($\in$ TK). These circles are arranged in a sequential order ($\in$ LS), and lines connect related chapters ($\in$ LK). This visualization is [LK·TK·LS]-based, and the technique is called the arc diagram. If we want to show groupings of items as well, we can add another pattern to the above blending: [LK·TK·LS·GR]. Using the above example, we can group some chapters together (see Figure 5.21b).

Now, consider subway and bus route maps. We have already seen an example of the Man-hattan subway map. Here we want to show a different design that uses a different blending. Recall that the first map we saw was [TR·LS·TK]-based. Figure 5.21c shows a generic subway map with station names, other labels, and routes. In this visualization, the designer has connected stations ($\in$ TK) to each other using lines ($\in$ LK). Stations that are linked form several successive ordered lists ($\in$ LS), with start and end stations that form different routes or linked-lists. Hence, this type of map is [LK-LS-TK]-based. It is interesting to note that this subway map and the first arc diagram are based on the blending of the same patterns.

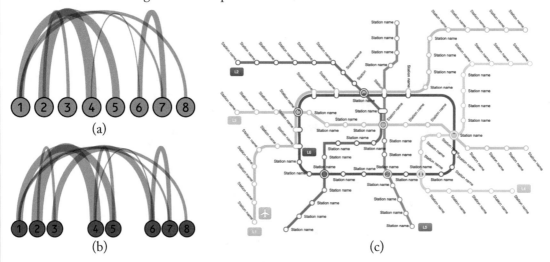

Figure 5.21: Examples of visualizations based on the Link pattern. (c) Courtesy of Edraw.

If our information items have clearly demarcated distances from each other, rather than using List, we can use an axis to give us a frame of reference: [CR•LK•TK]. Figure 5.22a is a representation, composed of a circular coordinate structure ($\in$ CR), marked using a set of rational numbers ($\in$ TK), and the lines connecting these ($\in$ LK). This visualization, called Farey diagram, is for exploring relationships among a set of rational numbers using geometry. Now assume that a designer needs to create a visualization that has cellular structures in which information items are placed. Additionally, these cells (or the items within them) have certain relationships with each other. One possible visualization is the one shown in Figure 5.22b. This representation uses Venn diagrams ($\in$ CL) for distributing the items ($\in$ TK) and links them together through connecting lines ($\in$ LK). This linked Venn diagrams technique is [LK•CL•TK]-based. Indeed, the NodeTrix technique (Henry et al., 2007) is also [LK•CL•TK]-based (see Figure 5.22c). This visualization technique combines two types of representations: matrix and node-link, where matrices are [CL•TK]-based, and lines connecting them are LK-based.

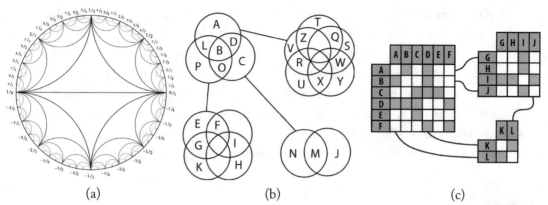

(a)	(b)	(c)

Figure 5.22: Examples of visualizations based on the Link pattern. (a) Courtesy of Allen Hatcher.

Blendings of List (LS)

Figure 5.23 shows how Apple's patent portfolio has evolved in yearly increments from 1996 to 2002. The visualization encodes the evolution of portfolios using a sequence ($\in$ LS) of treemaps, where each treemap ($\in$ [HR·CL·TK]) encodes the portfolio of a particular year. We can see that the biggest portfolio was in 1998. Colors encode some characteristics of the patents in a spectral fashion ($\in$ SP). Hence, the whole visualization is [LS·HR·CL·TK·SP]-based.

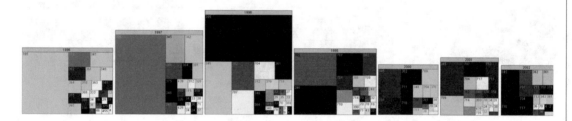

Figure 5.23: Example of a [LS·HR·CL·TK·SP]-based visualization. From Kutz (2004).

Figure 5.24a shows a diagram of the length of winter in four different cities during a four-year period. The visualization is composed of a polar coordinate system ($\in$ CR), where time is encoded using the concentric circles (increasing from center to outer rings). The arcs ($\in$ TK) in the rings of the polar coordinate system encode the length of winter for different years. Dots ($\in$ TK) on the arcs encode the midpoint of winter for each year. Midpoints are connected ($\in$ LK) together to encode changing trends in weather patterns for each city. Finally, the visualizations for the cities are placed sequentially ($\in$ LS). Hence, the whole visualization is [LS·CR·TK·LK]-based.

The visualization in Figure 5.24b depicts how gestures can be used to write a sentence using two different keyboards. The visualization consists of three sub-visualizations in the form of three vertical lists parallel to each other. The sub-visualization in the middle shows a sentence consisting of four successive (∈ LS) words (∈ TK) that need to be written using keyboard gestures. The other two sub-visualizations are composed of two vertical sequences (∈ LS) of small cell-like keyboards (∈ CL), with letters (∈ TK) placed on the keys. The gestures for writing the words are shown as continuous broken lines (∈ FS) that connect (∈ LK) the letters (∈ TK) and are overlaid on top of the cellular structure. Hence, the whole visualization is [LS•CL•TK•FS•LK]-based.

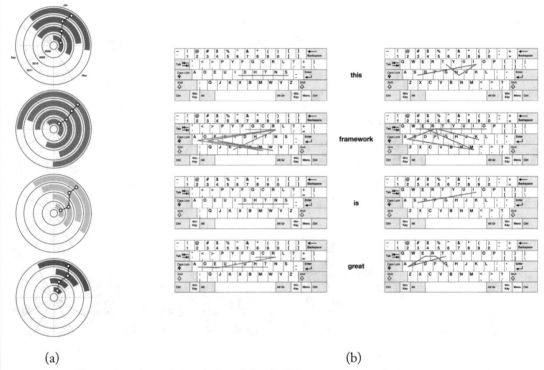

(a)   (b)

Figure 5.24: Examples of visualizations based on the List pattern.

Blendings of Hierarchy (HR)

Consider a node-link visualization technique. This technique is usually derived from a blending of Hierarchy, Token, and Link patterns. However, rather than having simple Token-based nodes, we can use other structures. Figure 5.25a shows an example of a visualization that is based on [HR•LS•LK•TK], where the nodes are [LS•TK]-based. The visualization represents a B+Tree.

The visualization in Figure 5.25b shows the evolutionary branching of different species over time, the domain of life to which the species belong, and the size of the genome of each species.

The inner tree-like component shows evolutionary branching from the last common ancestor of all the species, moving from the center to the circumference (∈ [HR•BR]). The species themselves are encoded at the inner circumference of the circle (∈ TK), and are placed together (∈ GR) to encode the three domains of life: eukaryota (animals, plants, and fungi), bacteria, and archaea. The bars at the outer circumference of the circle (∈ TK) encode the size of the genome of each species. Hence, the visualization is [HR•BR•GR•TK]-based.

The visualization in Figure 5.25c shows a set of entities (A to J) and some attributes that they share (1 to 4). The table-like structure is [CL•CR•TK]-based: representations encoding values (∈ TK) are placed in a partitioned structure (∈ CL), and a frame of reference is created by putting labels along its horizontal and vertical axes (∈ CR). Thus, visualizations can be referenced by their locations within the structure (e.g., A2, D4, etc.). The values within the cells are encoded in a spectral fashion to show their variation (∈ SP). To show the hierarchical nature of the entities, a dendrogram-style structure is placed on the left side that encodes the parent-child, multi-level organization (∈ [HR•BR]). The overall visualization is [CL•CR•SP•TK•HR•BR]-based.

Finally, Figure 5.25d shows the breakdown of a book (e.g., university textbook) and its intended use pattern. The visualization is composed of a circle with cells encoding the different chapters and sections of the book (∈ [HR•CL]), which are all labeled (∈ TK). Two types of relationships are encoded: (1) what chapters and sections refer to each other; and (2) sequential order in which chapters and sections should be read. The first type of relationship is encoded using dashed arrows that link those items together (∈ [LK•TK]). The second type is encoded in two ways: by placing consecutive sections in each chapter in a circular-like order (∈ LS), and by creating two linked lists of sequential sections and chapters (∈ [LK•LS])—one list for novice readers and the other for experts. Hence, the overall visualization is [HR•CL•LK•LS•TK]-based.

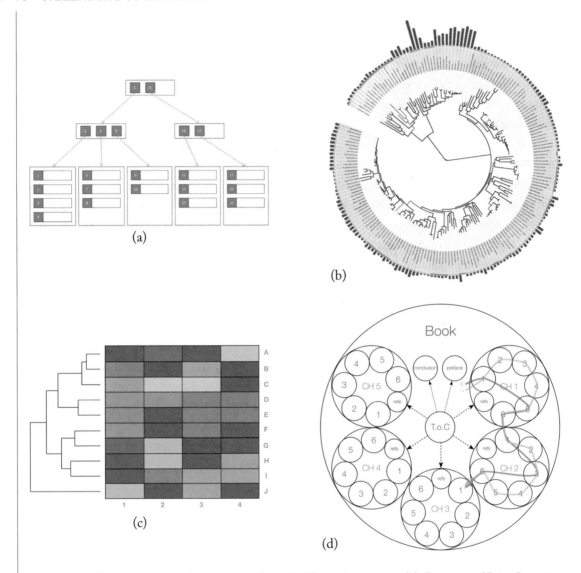

Figure 5.25: Examples of visualizations based on the Hierarchy pattern. (b) Courtesy of Ivica Letunić, author of iTol http://itol.emble.de. (d) Adapted from Iliinsky (2003).

Blendings of Spectrum (SP)

We have already seen a number of examples of how a designer can blend the Spectrum pattern with other patterns. In this section, we show two more interesting examples. Figure 5.26a shows a representation in which a designer has visualized harmonic functions that depict high magnetic fields in magnetic resonance spectroscopy. To do so, the designer represents the fields on two objects, a sphere and a cylinder section (∈ TK). Each instance from L to R represents a different second-order spherical harmonic function (X2-Y2, ZX, Z2, ZY, and XY), for the sphere (row A) and for the cylinder (row B). The variation in each function is shown in a spectral fashion, using different hues to encode values of the field (∈ SP). The fields have been normalized to a uniform minimum-to-maximum range of –1 to 1, which is encoded in a coordinate structure to provide a frame of reference (∈ CR). Thus, the visualization is [CR•FS•TK]-based. Figure 5.26b shows the visualization of the "thinking" of a chess game engine, where the chess board is encoded using a cellular partitioned structure (∈ CL) and chess pieces are encoded using symbols (∈ TK)—both black and white. The engine's thinking of possible moves from one location to another location are encoded using arcs (∈ LK). Black's and white's possible moves are encoded using two colors. The degrees of strength of possible moves are encoded in a spectral fashion (∈ SP)—the brighter the arc, the better the move. Hence, the visualization is [CL•TK•LK•SP]-based. Unlike the chess board visualization we saw before, the indexical location of chess pieces is not encoded explicitly, therefore the CR pattern is not part of this blend.

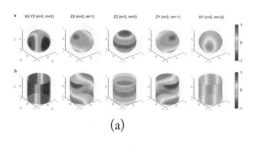

(a)

(b)

Figure 5.26: Examples of visualizations based on the Spectrum pattern. (a) From Juchem et al. (2006). (b) Courtesy of Martin Wattenberg.

Blendings of Stack (ST)

A visualization technique that is based on this pattern is the stacked graph that has several variations. Figure 5.27a shows a stack area chart representing the market share of five popular web browsers (i.e., IE, Chrome, Firefox, Safari, and Opera) over a certain year. This visualization is composed of a 2D coordinate system (∈ CR), a number of areas that are stacked on top of each other

($\in$ ST), where the areas represent a set of data points that have been fused together into integrated, continuous wholes ($\in$ FS). Hence, the visualization is [ST•CR•FS]-based. Another visualization technique that is [ST•CR•FS]-based is ThemeRiver (Havre et al., 2000), sometimes also called streamgraph. Figure 5.27b shows an example of this type of graph, composed of similar structures as those used in the previous example.

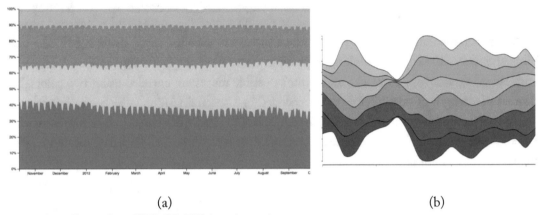

(a)   (b)

Figure 5.27: Examples of [ST•CR•FS]-based visualizations.

Another interesting example of the use of the Stack pattern can be seen in Figure 5.28. This visualization shows the polarization of U.S. political parties and senate networks. The horizontal axis ($\in$ CR) organizes representations with respect to time, increasing from L to R. Two vertical axes are employed and provide a frame of reference for the polarization modularity of each party ($\in$ CR). The coordinate system for the Democrats is stacked on top ($\in$ ST) of the one for the Republicans in a mirrored fashion, starting from the center and moving outward, up and down. The large dots ($\in$ TK) encode the party position (blue for Democrat, red for Republican) with respect to the modularity score for a two-year period. By mirroring the coordinate systems, increased distance between the large dots encodes increased polarity within the senate. Thus, we can see party polarization started to increase significantly at the beginning of the first Clinton presidency. Small dots encode individual senators ($\in$ TK), and links connect ($\in$ LK) senators to themselves at different points in time. Their position on the vertical axis is based on the balance of their votes with respect to the party (large dot). We can see that in recent years it has become rare for senators to occupy the middle ground, suggesting a lack of partisan cooperation. Saturation of the blue and the red colors is used to encode the variability and degrees of vote similarity ($\in$ SP). Finally, a temporal sequence of presidents ($\in$ LS) is seen at the bottom of the visualization. Hence, the visualization is [ST•CR•TK•LK•SP•LS]-based.

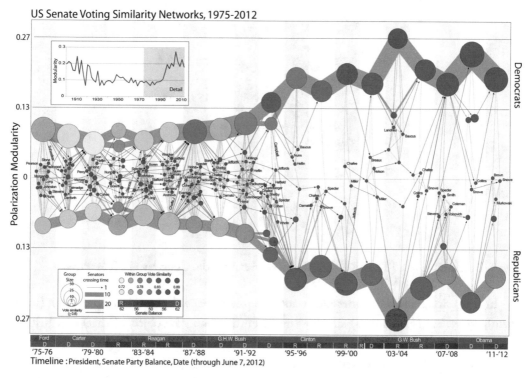

Figure 5.28: Example of a [ST•CR•TK•LK•SP]-based visualization. From Moody and Mucha (2013).

5.3   DESIGN AND STRUCTURAL ANALYSIS

In this section, we provide a few examples of interesting visualizations, and how the pattern language can not only help with their design, but also their structural and compositional analysis.

Our first example is from Physics. Consider an equipotential electromagnetic field as our information space. Some of the information items that need to be represented include: electric dipoles, a set of co-occurring electric fields with different boundaries, and equipotential lines that converge and diverge from the poles and are always moving perpendicular to the electric fields. To design a representation for this subset of the items from the information space, a designer can create a visualization such as the one in Figure 5.29. This visualization is composed of the positive and negative poles ($\in$ TK), the lines that converge and diverge from these poles ($\in$ BR), the arrows that show the direction of the flow of these equipotential lines ($\in$ TK), the two groups ($\in$ GR) of concentric ovals ($\in$ AR) that encode the co-occurring fields and bounded areas. The electromagnetic lines move through the concentric ovals (i.e., electromagnetic fields). Hence, this visualization, even though it may look simple, is [TK•BR•GR•AR]-based.

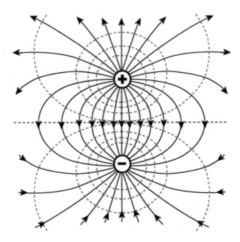

Figure 5.29: Example of a [TK•BR•GR•AR]-based visualization. Courtesy of HyperPhysics.

In some situations, a designer may want to represent *what* (categorical data), *when* (temporal data), and *where* (spatial data) simultaneously. Furthermore, the objects (what) may belong to different categorical sets and the spatial data (where) may have divisional regions. These features of the information space can suggest the use of several design patterns to start with: Token and Group for what; Area and Cell for where; and Coordinate for when. An innovative visualization that blends all these patterns plus Link is seen in Figure 5.30. It is composed of an outer circle that acts as a coordinate system, where different bio-hazard agents (categories) are grouped together. This part of the visualization is [TK•CR•GR]-based and encodes *what*. A set of concentric inner circles encodes time intervals (*when*), and are encoded using a polar coordinate system, where the inner circle shows the most recent time and its direction moves outward in a radial fashion. This part of the visualization is CR-based. A map of the U.S. showing different states ($\in$ [AR•CL]) is at the center of the visualization, representing *where*. Finally, relationships between what, when, and where are encoded using lines ($\in$ LK). This [TK•CR•GR•AR•CL•LK]-based representation is part of a set of visualizations for situational awareness.

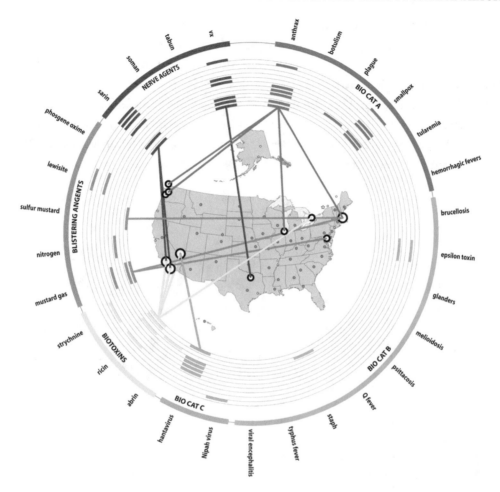

Figure 5.30: Example of a [TK•CR•GR•AR•CL•LK]-based visualization. From Livnat et al. (2005).

Consider an information space that consists of a set of related novels, where we want to communicate about some of the main characters, relationships, and events within the novels. Figure 5.31 shows a visualization that communicates about characters, relationships, and events within the five Game of Thrones novels. While examining the information space and determining how to generate a visual representation, the designer has decided that there are several interesting aspects that can be communicated. For instance, individual characters seem to exist within cliques that have mutual trust. Thus, individual characters are encoded by small circles ($\in$ TK), and are grouped according to clique by placing them close to one another within the inner dotted circles ($\in$ GR). The designer has also decided that the cliques could be grouped at a higher level. In other words, there is a hierarchical organization to the data. To encode this in the visualization, the cliques are placed as sub-systems of their super-system labels ($\in$ HR), which are shown on the circumference of the outer circle. For example, a number of characters are placed together to encode that they are part of the "Night Watch" clique. This clique, along with three others, is placed under the "Neutral" label on the outer circle, to encode the hierarchical nature of the groups. Finally, a specific type of relationship among individual characters (i.e., kills) is encoded using lines that are drawn between characters (∈ LK). The designer has also made use of some visual variables to encode properties of the characters (e.g., size of the circle encodes frequency of appearance, luminance encodes status, and so on). The varying status of the characters is encoded using color luminance (∈ SP). Thus, the whole visualization is based on [TK•GR•HR•LK•SP].

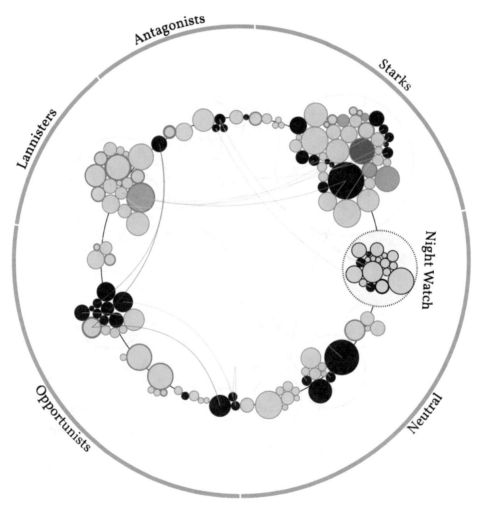

Figure 5.31: Example of a [TK•GR•HR•LK•SP]-based visualization. Courtesy of Jerome Cukier.

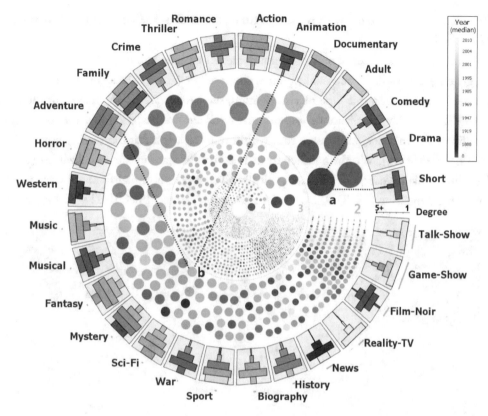

Figure 5.32: Example of a [CL•TK•CR•LS•SP]-based visualization. From Alsallakh et al. (2013).

Figure 5.32 shows movies and the genres to which they belong. The genres are arranged around the outside of the circle and compartmentalized in a cell-like fashion (∈ CL). Thus, each cell encodes a genre. The histogram sub-visualizations within the cells encode the relative number of movies within the genre, and are organized using the coordinate system from 1-5+ (∈ CR) to encode the number of other genres with which movies are shared. Within the talk-show genre, for example, we can see that most of the movies have only one genre; in the family genre, many of the movies are shared with two, three, or four other genres. Bubbles within the main circle (∈ TK) encode the existence of shared films between/among genres. Thus the bubbles each represent one or more films, where the size encodes the number of films. As the bubbles encode a shared relation among two, three, or four genres, they are placed into one of the corresponding concentric circles, so we see that comedy and short have many shared films with only each other, because the bubble is in the "2" circle (label a). We can see that animation and adventure have a fewer number of shared films with only each other (label b), since the bubble is smaller. Bubbles are arranged in a specific order from small to large in a clockwise fashion (∈ LS). Additionally, spectral coloring is used to encode variation of the movies with respect to time ($\in$ SP). Thus the visualization is derived from [CL•TK•CR•LS•SP].

Figure 5.33 shows a set of biological phenomena that have some relevance to an area of the human genome. The style of the visualization is commonly referred to as a genome browser in the bioinformatics domain. In this type of visualization, sub-visualizations are placed into horizontal tracks ($\in$ TR). The sub-visualizations within each track can be based on any number of different blendings. For instance, we can see that within one track is a stacked-area type visualization ($\in$ [ST•TK•CR•FS•SP]). Within another track is a type of stacked-bar visualization ($\in$ [ST•TK•CR]). The whole collection of tracks is typically given a chromosomal frame of reference along the horizontal dimension ($\in$ CR). The overall visualization is thus [TR•CR•ST•TK•SP•FS]-based.

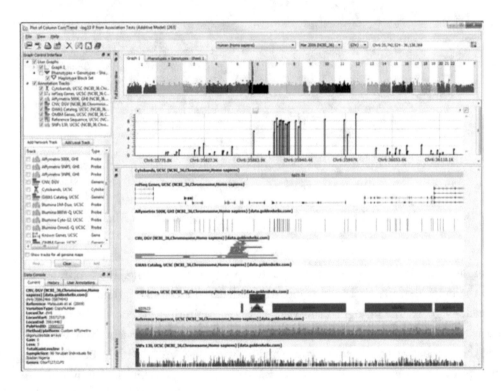

Figure 5.33: Example of a [TR•CR•ST•TK•SP•FS]-based visualization. Courtesy of Golden Helix © 2015.

Figure 5.34: Example of a [HR•TK•LK•CR•CL]-based visualization. From Burch and Diehl (2008).

Figure 5.34 shows a visualization that represents the results of soccer (football) matches among different nations. In the figure, we can see a node-link sub-visualization that encodes a hierarchy of the world with four levels from the middle of the circle to the outside (e.g., from World to Europe to Central Europe to Poland). Thus, each blue dot on the circumference of the circle encodes a country (i.e., a leaf node in the hierarchy). This sub-visualization is [HR•LK•TK]-based. With respect to the big circle in the middle, the concentric, nested circles act as a coordinate system and encode time intervals, increasing from the center to the outside (∈ CR). The concentric circles are partitioned into cells, with each cell within one of the concentric circles encoding a match

($\in$ CL). The color of the segment encodes the number of goals scored ($\in$ TK): black indicates 0, blue indicates 1–2, green indicates 3–4, yellow indicates 5–8, and red indicates more than 8. Finally, with regard to the large circle, pie-like slices of the circle encode different countries that are also encoded as a series of smaller circles around the larger circle. In the main circle, the segments encode "incoming" edges to the country that is at the edge of the circle. In the small circles for each country, the segments encode "outgoing" edges from that country. Thus, in the main circle, color is the number of "incoming" goals (i.e., goals scored against) ($\in$ TK). In the small circles, color encodes the number of "outgoing" goals (i.e., goals scored for) ($\in$ TK). To make a mental connection between incoming and outgoing, the viewer has to look at the corresponding "slice" in each circle. For example, if we look at the slice for Liechtenstein, we see that it has the only red incoming edge, meaning that more than eight goals were scored against it. To find out which team scored the goals, we have to look at the slice that represents Liechtenstein in the small circles—we see that Germany has the matching red outgoing edge. So we conclude that Germany scored more than eight goals against Liechtenstein on that occasion. The user can click to drill into that specific match and get details about it. We can also see general patterns in the information, such as: Austria is allowing more goals over time (more green near the outside); Suriname has played only a few games, all against French Guiana; very few games have been played between Central Europe and South America (top half of the circles are mostly empty for South American countries). A notable exception is Brazil vs. Germany. Thus, we can see that this complex visualization results from the blending of the following patterns: Hierarchy, Link, Token, Coordinate, and Cell.

CHAPTER 6

Human–Information Interaction

As was stated in Chapter 1, much of the activity in which people are engaged involves using and working with visual representations of information. In all but simple situations, being able to interact with the representations significantly enhances the efficacy of the activity. This is especially true in the context of big data and large information spaces, which are becoming increasingly characteristic of the activities in which people are engaged. Thus, in the context of visualization design, making visualizations interactive is imperative for supporting complex activities. In other words, visualization tools should be designed such that users can engage in an ongoing discourse with the underlying information as they carry out different tasks and activities. This discourse is referred to using different labels, such as human-information interaction, human-information discourse, or human-data interaction. In complex contexts, users can rarely accomplish their goals through visual tasks with a single visualization alone. Rather, users need to see multiple visualizations that provide different perspectives on the underlying information, and be able to perform varied interactions to support different interrelated and interdependent tasks. As a result, designers need to think from the outset about providing users with multiple visualizations, each with its own appropriate interactions, to support users in effectively engaging with their information. Even though this book is geared toward designing visual representations, and a detailed treatment of human-information interaction is outside its scope, we have included this brief chapter due to the critical role that interaction plays in visualization use and design.

In Section 6.1 we discuss the idea of humans and visualization tools functioning as joint cognitive systems. We briefly categorize such systems into different spaces that can be used in design thinking. In Section 6.2 we discuss the relationships between activities, tasks, and interactions in the context of complex cognitive activities. In Section 6.3 we discuss interaction design, where we make a distinction between interaction and interactivity. We present some interaction patterns that designers can use in tandem with this visualization framework to generate a variety of interactive visualizations and highlight the importance of ontological and operational aspects of user-visualization interaction in design thinking.

6.1 HUMAN-VISUALIZATION SYSTEM

When using visualization tools to support the performance of complex cognitive activities, the user(s) and the tool are coupled together such that they form a joint cognitive system (Fisher et al., 2011; Liu et al., 2008; Parsons and Sedig, 2013c; Sedig and Parsons, 2013). In such a system, cognitive processing is distributed across the internal (mental) representations and processes of a user, and the external (visual) representations and processes in the environment (Hollan et al., 2000; Zhang and Norman, 1994). This view of cognition, which has become widely accepted in recent decades, is known as distributed cognition (see Hutchins, 1995; Salomon, 1993). Research throughout the past few decades has demonstrated that interactive computational tools—when designed well—can become active partners in cognition, distributing the requisite load of cognitive processing (Jonassen, 1995; Lajoie and Derry, 1993; Liu et al., 2008; Salomon et al., 1991; Sedig et al., 2001). We mention distributed cognition only briefly here to situate the reader's thinking, and to suggest a conceptual lens for design. For more detail, we refer the reader to the following sources: Hollan et al., 2000; Kirsh, 1997, 2005, 2010; Liu et al., 2008; Nardi, 1996; Parsons and Sedig, 2013c; Salomon, 1993; Sedig and Parsons, 2013; Zhang and Patel, 2006.

To think systematically about design issues in the context of joint cognitive systems (i.e., the user and the visualization tool), it can be useful to think about the different components of the system and their requisite design considerations. In previous work, we have categorized this system into five spaces: information, computing, representation, interaction, and mental space (see Sedig and Parsons, 2013 for a detailed discussion). While information and representation spaces have been described in detail in Chapter 3, the others can be described very briefly as follows: *computing space* is the place where data is processed, stored, and prepared. This space may involve data cleaning, fusion, filtering and other pre-processing procedures, as well as data mining, transformation, and analytics procedures. Design situations with static media (e.g., print) do not have this space. Users perform mental operations within the *mental space*—the place in which internal mental events and operations take place (e.g., memory encoding, storage, and retrieval; apprehension; judgment), and in which mental structures exist (e.g., mental models and schemas). The *interaction space* consists of actions that users perform and reactions that occur within the representation and computing spaces. All of these spaces together form the joint cognitive system, across which information processing is distributed in complex activities (Parsons and Sedig, 2013c). It is important to note that these spaces do not together constitute the full visualization *design space*. Rather, they are meant to function as a conceptual lens for analyzing the human-visualization system, and not for capturing all relevant components of the design space. In later chapters, these spaces will be revisited and expanded in the context of design. Figure 6.1 depicts these spaces within the human-visualization system. As we have discussed in Chapter 3, both the representation space and the information space can be conceptualized as systems. Figure 6.1 depicts this conceptual lens using a multi-level abstract representation of a system.

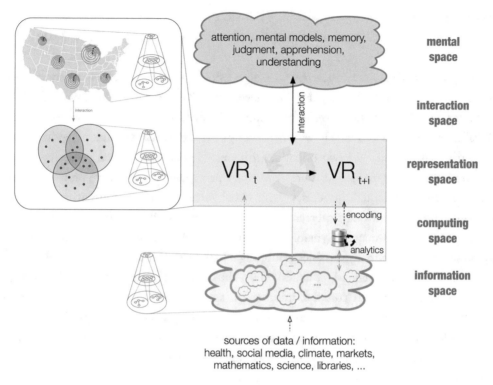

Figure 6.1: A division of the human-visualization system into multiple spaces. VR stands for visual representation. Adapted from Sedig and Parsons (2013).

6.2 ACTIVITY, TASK, AND INTERACTION: A CHARACTERIZATION

Although used extensively in the visualization literature, there are currently no precise definitions of the terms *task* and *activity* that are widely agreed upon (Rind et al., 2015). As in other aspects of visualization design, there are multiple levels of abstraction that can be characterized when defining tasks and activities. Similar to Norman (2005) and others (e.g., Cooper et al., 2014), we view these in a hierarchical fashion, and consider an activity to be at the highest level, with the activity being made up of a number of tasks that are carried out to achieve the overall goal of the activity. In the visualization literature, tasks have been examined at multiple levels of granularity. The more abstract characterizations tend to refer to generalized and domain-independent tasks. At this level of abstraction, tasks are difficult to characterize in detail. As tasks become more domain-specific and concrete, they become easier to characterize, but more difficult to generalize. Engaging in a detailed examination of tasks is outside the scope of the visualization design framework presented in this book. Many other contributions have examined the concept of tasks in the visualization literature to which the reader can refer—e.g., Amar et al., 2005; Brehmer and Munzner, 2013; Rind et al., 2015; Schulz et al., 2013; Sedig and Sumner, 2006; Shneiderman, 1996; Tominski, 2015; Zhou and Feiner, 1998.

In this book, we are not concerned with all possible activities; rather, we are concerned with cognitive activities—those that involve high-level cognitive processes—examples of which are sensemaking, decision-making, learning, planning, problem solving, and analytical reasoning. More specifically, we are concerned largely with *complex cognitive activities*.[9] Complex cognitive activities are activities that have at least two essential attributes: (1) complex mental processes—requiring the combination and interaction of more elementary perceptual and attentive processes as well as higher-level controlled thought processes—e.g., visual perception, focused attention, memory processes, mental models, and analytical reasoning; and (2) complex external conditions—e.g., external processes and structures may be dynamic, non-linear, non-deterministic, characterized by uncertainty, noise, and/or with many variables exhibiting a high level of interaction and interdependence (Knauff and Wolf, 2010; Schmid, et al., 2011). In the context of visualization, complex cognitive activities are performed with complex information spaces (e.g., large, dynamic, interconnected, interdependent, heterogeneous information spaces). Examples of complex cognitive activities are analyzing large collections of medical documents to make clinical diagnoses, and exploring multiple heterogeneous datasets to make sense of global climate change patterns. To design effectively in this context, designers need adequate support structures to help them think systematically about relevant design issues. Design guidelines for supporting simpler tasks are not entirely sufficient to guide design in the context of complex activities. For example, the well-known mantra "overview first, zoom and filter, details-on-demand" (Shneiderman, 1996)—while still useful for guiding the design of certain aspects of a visualization tool—is likely not appropriate guidance for all aspects of the tool. In the age of big data and increasingly large information spaces made of federations of heterogeneous datasets, designers need frameworks to help them quickly and appropriately generate a multiplicity of types of interactive visualizations to help users work effectively in complex situations.

Figure 6.2 depicts the hierarchical nature of a complex cognitive activity. An activity is made up of tasks. Tasks involve some cognitive processing—e.g., categorizing items that are stored in short-term memory. Tasks typically also require visual processing—e.g., locating items in the representation space. Furthermore, tasks often have an interactive aspect to them—i.e., the user interacts with visualizations while performing tasks. These are often all coordinated to carry out a single overall task—e.g., interactively rearranging visualizations, visually locating items, and mentally categorizing them. Thus, when conceptualizing tasks, they can be thought of as having three aspects: cognitive, interactive, and visual. Different tasks require different distributions of visual, cognitive, and interactive processing. The distribution of processing is not only context- or domain-dependent, but also user-dependent—since some users may be able to perform a task with higher or lower

---

[9] We use the terms *complex cognitive activity* and *complex activity* interchangeably throughout the book.

numbers of external interactive structures, depending on their prior knowledge, domain experience, level of expertise, and/or visualization literacy. While cognitive and visual tasks always take place when users are working with visualizations, interactive tasks do not take place if visualizations are static. Hence, they are encoded as optional in Figure 6.2. Interactive tasks operate in the interaction space, and can be viewed at different levels of granularity—from abstract interaction patterns (e.g., drilling), to interaction techniques (e.g., details-on-demand), to physical events (e.g., mouse clicks). Due to space limitations, we do not wish to delve into a deep examination of these different levels of interaction; the interested reader can refer to (Sedig and Parsons, 2013) and (Sedig et al., 2012) for more information about the conceptualization of these levels. Suffice it to say that, similar to visualization design, there are different levels of abstraction involved here that the designer should consider—everything other than physical events can be considered as conceptual design.

Visual tasks can be broken down into low-level and high-level tasks. Low-level visual tasks are often executed pre-attentively. However, without delving into the minutiae of pre-attentive versus attentive processing, we simply use the terms low- and high-level. Examples of low-level visual tasks are: detecting the presence or absence of a unique visual feature in a visual scene, discriminating among a set of visual objects, detecting a boundary between groups of elements, comparing the lengths of two lines, and comparing the areas of multiple circles. Examples of high-level visual tasks are: scanning the representation space to locate a particular object, tracing routes or paths between and among objects, and categorizing objects according to their visual features. The main difference between these levels is that low-level visual tasks are performed automatically, without focused attention, and/or are performed in a very short period of time. High-level visual tasks require focused attention and take more time (e.g., many seconds) to perform. As discussed in Chapter 3, the effects of visual variables on visual tasks occur primarily at the low level, and are especially accentuated when the time available to perform the tasks is very short (Carswell, 1992).

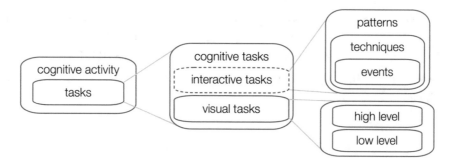

Figure 6.2: Breakdown of a cognitive activity. An activity is made of tasks, which can be cognitive, interactive (optional), and/or visual. Interactive tasks can be viewed at the level of interaction patterns, techniques, or events. Visual tasks can be broken down into high-level and low-level tasks.

In most complex situations, activities are usually composed of sub-activities, and tasks are composed of sub-tasks. For example, consider an activity such as *making sense* of an information space consisting of millions of legal documents. To perform the activity, a user may need to perform an *analytical reasoning* sub-activity to systematically break down the entities in the space and reason about their relationships. To carry out this sub-activity, the user may need to perform a series of tasks, such as *browsing* through the documents from a particular decade, *categorizing* them according to a major theme, *ranking* them according to their importance, and so on. These tasks may be repeated throughout the activity, sometimes being sub-tasks of other tasks. Each of these tasks may then require sub-tasks, and so on. Each task would likely involve the performance of a number of interactions. For instance, to categorize the documents, the user may need to *filter* them according to a particular attribute or *drill* into a particular document to access more information. To complete any one of these actions, a number of micro-level events such as mouse clicks, finger swipes, or keystrokes may be required.

During design, the relationships among activities, tasks, and interactions can be viewed from a bidirectional perspective: bottom-up and/or top-down. From a bottom-up view, the performance of a task gradually emerges over time through a sequence of interactions that users perform with visual representations. Likewise, an activity emerges over time through a sequence of tasks that users carry out. From a top-down view, activities can be broken down into a sequence of tasks, and tasks can be broken down into a sequence of interactions that are performed at the level of the representation space. As a designer, it may be beneficial to think in both a bottom-up and in a top-down fashion, and to ask questions from both perspectives—e.g., "what tasks will users be able to accomplish with the interactions and visualizations that are provided?," or "what tasks should my visualization tool support so that users can effectively carry out their activities?" Figure 6.3 depicts the hierarchical nature of activities, tasks, and interactions as they are performed over time. As described above, an overall cognitive activity is composed of sub-activities; sub-activities are composed of tasks; and, tasks are composed of sub-tasks. The sub-tasks are carried out through a series of interactions. At the interaction level, a user performs a series of actions on visualizations (A1, A2, …), which lead to a series of reactions in the representation space (R1, R2, …). All the while, the user is perceiving what is happening in the representation space (labeled at different points in time, P1, P2, P3, …).

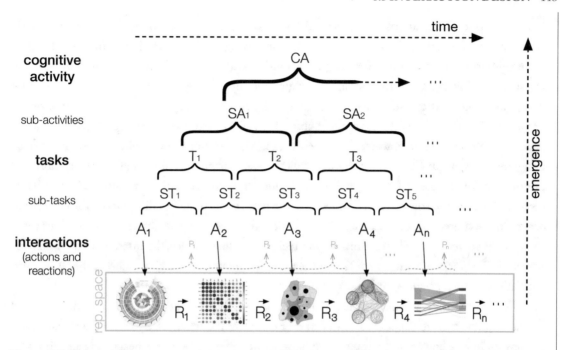

Figure 6.3: Relationships among activities, tasks, and interactions. Top-down view: cognitive activity is made up of sub-activities, tasks, sub-tasks, interactions, and perceptions. Bottom-up view: cognitive activity emerges over time, through performance of interactions and tasks.

6.3   INTERACTION DESIGN

Since tasks and activities are emergent phenomena, they cannot be designed directly. Interactions, however, can be designed directly. Tasks and activities emerge from a sequence of interactions that are performed; thus, it is important for designers to have a good understanding of how interactions can and should be designed to best support the performance of tasks and activities. In this section we present two aspects of interaction design for visualizations: interaction and interactivity.

6.3.1   INTERACTION

When viewing visualizations, users always perform tasks and activities—whether the visualization is interactive or not. For example, users can solve problems, plan, locate items, compare them, correlate them, and so on, all by looking at a static visualization. When doing so, the necessary information processing involved in carrying out the task or activity is distributed across the differ-ent spaces in the human-visualization system. Depending on the complexity of the activity, and a number of other factors, different distributions are more desirable—e.g., placing more load on the computing space and removing it from the mental space. One way to effectively reduce the mental load is to make visualizations interactive. Although our framework supports the creation of static visualizations, we suggest its main utility is in supporting visualization design in the context of human-information interaction for complex activities. Although we do not delve into great detail about interaction design in this book, we have previously developed a framework for interaction design in the context of complex activities (see Sedig and Parsons, 2013). One component of the framework is a catalog of 32 fundamental epistemic action patterns—where epistemic actions are actions taken by users to transform the world (i.e., visualizations) to facilitate mental information-processing needs (Kirsh and Maglio, 1994). These epistemic actions that a visualization tool can provide for the user are listed in Table 6.1. Similar to the visualization patterns in this book, these interaction patterns can be considered as the basic conceptual building blocks for interaction design in the context of complex cognitive activities. These patterns are abstractions—each of them can be instantiated using different interaction techniques; for instance, the interaction pattern *drilling* can be instantiated using the following techniques: mouse-over, right-clicks, spatial proximity, semantic zooming, gestures, and digital probes.

When the right interactions are chosen in design, they can be combined and incorporated in visualization tools to support users in carrying out complex activities. By using the visualization framework in this book and the interaction patterns in Table 6.1 together, designers can think systematically about generating multiple highly interactive visualizations, enabling users to engage in deep discourse with the underlying data and information.

Table 6.1: Epistemic interaction patterns from Sedig and Parsons (2013)

Action	*Characterization*: acting upon visualizations to …
Annotating	augment them with additional visual marks and coding schemes, as personal meta-information
Arranging	change their ordering, either spatially or temporally
Assigning	bind a feature or value to them (e.g., meaning, function, or behavior)
Blending	fuse them together such that they become one indivisible, single, new visualization
Cloning	create multiple identical copies
Comparing	determine degree of similarity or difference between them
Drilling	bring out, make available, and display interior, deep information
Filtering	display a subset of their elements according to certain criteria
Measuring	quantify some items (e.g., area, length, mass, temperature, and speed)
Navigating	move on, through, and/or around them
Scoping	dynamically work forward and backward to view compositional development and growth

Searching	seek out the existence of or locate position of specific items, relationships, or structures
Selecting	focus on or choose them, either as an individual or as a group
Sharing	make them accessible to other people
Transforming	change their geometric form
Translating	convert them into alternative informationally- or conceptually-equivalent forms
Accelerating/ Decelerating	increase or decrease speed of movement of their constituent components
Animating/ Freezing	generate or stop motion in their constituent components
Collapsing/ Expanding	fold in or compact them, or conversely, fold them out or make them diffuse
Composing/ Decomposing	assemble them and join them together to create a new, whole visualization, or conversely, break whole entities up into separate, constituent components
Gathering/ Discarding	gather them into a collection, or conversely, throw them away completely
Inserting/ Removing	interject new visualizations into them, or conversely, get rid of their unwanted or unnecessary portions
Linking/ Unlinking	establish a relationship or association between them, or conversely, dissociate them and disconnect their relationships
Storing/ Retrieving	put them aside for later use, or conversely, bring stored visualizations back into usage

6.3.2 INTERACTIVITY

Much of the discussion on interaction design in the visualization literature focuses on the *ontological* aspects of interaction—e.g., what interactions exist, at what level of abstraction interactions can be identified and characterized, what properties of individual interactions can be identified, and how interactions can be categorized. Much less focus has been placed on the *operational* aspects of interaction—e.g., how interactions can and should be put into operation. For example, consider filtering—an interaction pattern from Table 6.1. An ontological characterization of filtering is as follows: when this interaction pattern is implemented, a user can act upon the representation space to show or hide a subset of the information in a visualization according to some criteria. In terms of how a filtering interaction is operationalized, there are a number of possibilities—e.g., the user could issue a linguistic command, either by speaking or by typing on a keyboard; the user could click and drag a slider; or the user could act directly upon the visualization (i.e., Direct Manipulation). In each of these cases, the ontological aspect of the interaction is consistent: the user interacts with the visualization to show or hide some information. However, in each case the operational aspect is different. Although there is no doubt about the relevance and importance of the ontological aspects of interaction for designing visualizations, designers also need to consider the operational aspects of interaction in their design processes. For years there has been accumulating research evidence that different operational aspects of interaction can significantly impact the cognitive processes of users and, hence, the performance of higher-level tasks (e.g., Corbett and Anderson, 2001; Hutchins et al., 1985; Kieras et al., 2001; Liang et al., 2010; Liu and Heer, 2014; Sedig and Haworth, 2014; Sedig and Liang, 2006; Sedig et al., 2001; Sedig et al., 2005b; Shneiderman, 1982; Svendsen, 1991; Trudel and Payne, 1995). In previous work, we have discussed these issues in the context of *interactivity*. The suffix "ity" can be used to form nouns that denote the quality or condition of something; therefore, in this context, while interaction refers to reciprocal action—that is, action and reaction—interactivity refers to the *quality or condition of interaction*. In a previous contribution, we have presented a framework for analysis of interactivity of visualizations, both at the micro and macro levels of design (see Sedig et al., 2013). Interactivity at the micro level emerges from the structural elements of individual interactions. Interactivity at the macro level emerges from the combination, sequencing, and aggregate properties and relationships of multiple interactions as a user performs tasks and activities. Table 6.2 presents a summary of the micro-level interactivity elements and their possible operational forms. Once again, it is important to note that, similar to visualization design, these elements, though very close to concrete design at the spatio-temporal level of events that operate on visualizations, are still conceptual constructs to help interaction design thinking. Table 6.3 presents a summary of the macro-level interactivity factors.

Table 6.2: Micro-level interactivity elements

Component	Element	Concern	Forms
action	agency	metaphoric agency through which action is expressed	verbal, manual, pedal, aerial
	flow	parsing of action in time	discrete, continuous
	focus	focal point of action	direct, indirect
	granularity	constituent steps of action	atomic, composite
	presence	existence and advertisement of action	explicit, implicit
	timing	time available to user to compose and/or commit action	user-paced, tool-paced
reaction	activation	point at which reaction begins	immediate, delayed, on-demand
	context	context in which visualization exists as interface reaches equilibrium	changed, unchanged
	flow	parsing of reaction in time	discrete, continuous
	transition	presentation of change	stacked, distributed
	spread	spread of effect that action causes	self-contained, propagated
	state	condition of visualizations once reaction process is complete	created, deleted, altered

Table 6.3: Macro-level interactivity factors

Factor	Concern
Diversity	number and diversity of interactions that are available to the user
Complementarity	harmonious and reciprocal relationships among interactions, and how well they work with and supplement each other
Fitness	appropriateness of interactions for the given visualizations, the tasks and the activity, and the user's needs and characteristics
Flexibility	range and availability of adjustability options
Genre	types of transactions that are available to the user—i.e., interactions through which the user makes exchanges with the visualizations

6.3.3 FURTHER SOURCES

By making the distinction between interaction and interactivity above, the ontological and operational aspects of interaction can be clearly separated and discussed coherently in visualization design contexts. Designers can think, in a clear and informed manner, about which interactions they want to provide to users, and how their operational forms can affect cognitive processes and the overall quality of the human-information discourse.

Since our main focus here is on the design of visual representations, it is beyond the scope of this book to have an in-depth discussion of interaction design. Hence, our treatment of this topic is cursory. However, we would like to emphasize that effective design of interactive visualizations depends on the designer carefully considering the ontological as well as the operational aspects of interaction—namely, interaction and interactivity—as they are both essential elements of design of visualizations for human-information interaction in the context of complex cognitive activities. We strongly encourage readers to consult other relevant sources: Elmqvist et al., 2011; Green and Fisher, 2011; Heer and Shneiderman, 2012; Kirsh, 1997, 2013; Pike et al., 2009; Roth, 2013; Tominski, 2015; Yi et al., 2007. The authors' own thinking regarding interaction and interactivity design for visualizations can be found in the following articles: Morey et al., 2001; Sedig and Morey, 2002; Morey and Sedig, 2004; Sedig and Sumner, 2006; Sedig and Liang, 2008; Liang et al., 2010; Liang and Sedig, 2010; Sedig et al., 2012; Sedig and Parsons, 2013; Sedig et al., 2013; Parsons and Sedig, 2013b; Parsons and Sedig, 2013c.

CHAPTER   7

Design Process

In this chapter we present a simple process for designing visualizations for human-information interaction. This process can be considered as the capstone of the framework. The description of this element of the framework is brief and covers the design process at a high level. The proposed process is intended specifically for designing visualizations. Thus, it should complement, and not replace, other relevant design processes, principles, heuristics, frameworks, and methodologies (see e.g., Carroll, 2000; Johnson, 2013; Lidwell et al., 2010; Munzner, 2009; Nelson and Stolterman, 2012; Norman, 1999, 2013; Norman and Draper, 1986; Reeves et al., 2002; Sedlmair et al., 2012; Shneiderman et al., 2009; Tidwell, 2005; Ware, 2012; Weinschenk, 2011). The next chapter presents the application of the visualization framework using a number of examples to illustrate how it can help with design thinking.

7.1   DESIGN STAGES

Figure 7.1 shows a cyclical diagram that represents the visualization design process. This process brings together and integrates the conceptual elements of the framework, which we have discussed throughout the book, into a coherent whole. The process is made of four major stages that the designer can follow. Each stage is described in more detail in the subsections below. Since many of the stages involve selecting from a space of possibilities, a series of accompanying tables are presented in the relevant subsections. These can be consulted by designers as needed.

Even though the design process is depicted in a sequential manner, it is important to note that, in real-world design scenarios, the process is usually iterative. This is depicted by the dashed line between Stages 4 and 1. In addition, at every stage, the designer can and is encouraged to go back to the previous stages to modify and improve the design. Moreover, although not depicted in the diagram, the designer can move freely among the stages, skipping one or more if needed, especially after fully going through the whole process once. Finally, for complex visualizations that have many sub-visualizations at different levels, the process can actually be repeated many times for each sub-visualization. Thus, the process can be nested within itself at many different levels.

One more important point that readers should keep in mind is that this process by no means advocates that exisiting visualization techniques should not be used, nor does it suggest that in certain design scenarios designers cannot simply use an existing technique such as a histogram, scatterplot, or treemap. This design process is primarily intended to promote systematicity and creativity in the context of generating novel visualizations, especially in the context of human-information interaction with complex information spaces, where users benefit from a multiplicity of interactive visualizations.

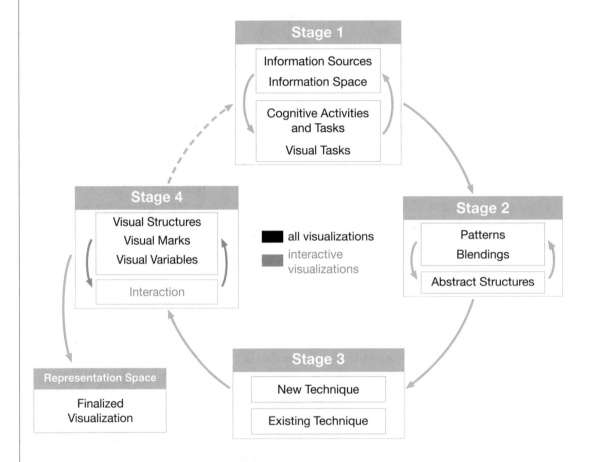

Figure 7.1: The visualization design process, comprising four main stages. Designers can traverse the stages many times, in an iterative fashion, until a visualization is finalized.

7.1.1 STAGE 1: INFORMATION SPACE AND TASK SPACE

This stage involves analyzing two spaces: information space and task space. Even though these are two separate spaces, in the process of their analysis, the designer should go back and forth between the two—that is, analysis of one affects how the other is analyzed.

The designer can start with the information space. Given the concepts discussed in Chapter 3, the designer may want to ask a series of questions about the information space, the answers to which have a direct effect on how to proceed. Table 7.1 can help with this process. Some of the questions that the designer can ask include:

- What are the sources of the information?

- Is this information space open or closed? Does it receive incoming new information?

- Does the information space have sub-spaces?

- Does the information space have many levels? If yes, what is the depth?

- Is the information space homogeneous or heterogeneous? (e.g., structured data, algorithmically generated data, textual documents, derived information, real-time data, etc.)

- Are there relations among the sub-spaces?

- What are the properties of the sub-spaces? (e.g., size, cardinality, dynamism, etc.)

- What are the typologies of the information items?

- At what level of granularity should the items be encoded?

- And so on.

For more on data and information space analysis, the interested reader can also refer to the following sources: Boisot and Canals, 2007; Card et al., 1999; Hansen and Johnson, 2005; Kreuseler and Schumann, 2002; MachEachren, 1995; Meadows, 2008; Munzner, 2015; Roth and Mattis, 1990; Shneiderman, 1996; Skyttner, 2005; Tominski, 2015; Tory and Möller, 2004; Ward et al., 2015; Ware, 2012; Wilkinson, 1999; Zhou and Feiner, 1998; Zins, 2007.

The next step in Stage 1 involves an analysis of the tasks and activities that users are going to perform with the given information space. This includes low-level visual tasks, higher-level cognitive tasks, and even complex cognitive activities of the users. Once again, given the concepts discussed in Chapter 6, the designer may want to analyze the task space to see how user goals and intentions should be supported. Table 7.1 can help with this process. Questions here revolve around both non-interactive visual tasks as well as some interactive ones. For instance, with regard to low-level visual tasks, if two of the expected user tasks are to visually "navigate a sequence of objects" and to "trace the boundary of a region," then these visual tasks will impact the selection of the patterns from Stage 2. If the designer knows that users will want to categorize, reorganize, and rank items, design decisions can be made accordingly in subsequent stages in the process. An analysis of this space may also lead to a re-analysis of the information space, for instance to change the level of granularity at which the information items are conceptualized. It should be noted that analysis of the task space can be carried out in a number of ways. There are many existing guidelines and methodologies related to task analysis. The interested reader can consult the following sources: Brehmer and Munzner, 2013; Diaper and Stanton, 2003; Hackos and Redish, 1998; Jonassen et al.,

1998; Pirolli and Card, 2005; Potter et al., 2000; Rind et al., 2015; Schraagen et al., 2000. Finally, in Stage 1, the designer should iteratively go back and forth between analyzing the information and the task spaces until she has a good sense of what is needed for Stage 2.

Table 7.1: A reference table for Stage 1 of the visualization design process: information space and task space

Information Space		Task Space	
Sources	Systems	Cognitive Activities and Tasks	Visual Tasks
Health	Super-systems	Analyzing	Associating
Education	Sub-systems	Decision making	Comparing
Sports	Co-systems	Diagnosing	Counting
Climate	Open system	Discovering	Detecting
Economy	Closed system	Experimenting	Discriminating
Crime	Entities	Exploring	Finding
Religion	Properties	Interpreting	Focusing
Science	Relations	Investigating	Hopping
Libraries	Items	Learning	Identifying
Demographics	Sub-spaces	Planning	Locating
Manufacturing	Levels	Predicting	Looking-only
Resources	Granularity	Problem solving	Observing
Entertainment	Dimentionality	Reasoning	Orienting
...	Data types	Sense making	Pointing
	Dynamism	Summarizing	Recognizing
	Homogeneity	Triaging	Scanning
	Heterogeneity	...	Tracing
	...		Walking
			...

7.1.2 STAGE 2: PATTERNS, BLENDINGS, AND ABSTRACT STRUCTURES

This stage consists of three steps: selection of patterns, blending of patterns, and consideration of abstract structures (see Table 7.2 for examples of these). These concepts were discussed in Chapters 3, 4, and 5. The first step involves thinking about and selecting the pattern(s) that are needed to map items from the information space to the representation space. The next step involves deciding how to blend the patterns that were selected in the first step. At this point, it is important for the designer to give thought to whether the selected patterns and their blendings provide good mappings from the information space to the representation space and can support the tasks of the user, which were analyzed in Stage 1. For instance, for the task of visually navigating a sequence of objects, the designer may consider the List pattern. Depending on what the term "objects" signifies, the designer can either select the Token pattern or a blending of other patterns. Step 3 in this stage involves a consideration of, and an analysis of, the abstract structures that can be derived from the patterns and their blendings, as well as their organizational affordances. For instance, if the Area and Cell patterns have been selected and blended, the designer may consider geographic and containment structures, whereas, a graph structure may not seem suitable. Two things are important here: (1) that these structures being considered are still abstract, but closer to concrete instantiations than the blendings themselves; and (2) that thinking about these structures can help the designer reconsider which patterns to select and blend. As a result, the designer should iteratively go back and forth between the steps in this stage until she has a good sense of how the patterns and structures can result in desired visualization techniques.

Table 7.2: A reference table for Stage 2 of the visualization design process: Patterns, blendings, and abstract structures

Patterns, Blendings, and Abstract Structures		
Patterns	**Pattern Blendings**	**Abstract Structures**
Primary:	[AR·BR·TK]	Areal structures
Fusion (FS)	[AR·CL·TK]	Cellular structures
Token (TK)	[BR·FS·CR·TK]	Containment structures
	[CC·CL·GR·TK]	Coordinate structures
Substrate:	[CL·HR·TK·SP]	Geographic structures
Area (AR)	[CL·CR·TK·BR]	Geometric structures
Cell (CL)	[CR·FS·TK·GR]	Graph structures
Coordinate (CR)	[HR·LS·LK·TK]	Hierarchical structures
Track (TR)	[LK·TK·LS·GR]	Tabular structures
	[LS·HR·CL·SP]	Topological structures
Relational:	[ST·CR·TK·LK]	Tree structures
Branch (BR	[TR·CR·SP·TK]	…
Cycle (CC)	[TR·LS·AR·SP]	
Group (GR)	[CL·TK·CR·LS·SP]	
Hierarchy (HR)	[TK·GR·HR·LK·SP]	
Link (LK)	[TK·CR·GR·AR·CL·LK]	
List (LS)	[TK·BR·GR·AR]	
Spectrum (SP)	…	
Stack (ST)		

7.1.3 STAGE 3: VISUALIZATION TECHNIQUES

Given the blending of the patterns and the consideration of possible structures in Stage 2, Stage 3 involves either the creation of a new visualization technique or the selection of an existing technique from the space of hundreds of techniques that have already been developed. We have already dicussed and demonstrated in Chapters 4 and 5 that the pattern language provides a great deal of combinatorial flexibility when thinking about design possibilities. A blending of patterns can inspire the designer to generate an innovative and novel visualization technique which can be adopted and developed at Stage 3. Alternatively, the blendings and structures in Stage 2 can narrow down the space of design possibilities and help the designer select an appropriate visualization technique from the space of existing techniques. For instance, having decided to blend the Coordinate and Token patterns, and having considered the use of coordinate and multi-dimensional glyph structures, the designer may select a scatterplot as a suitable technique.

Table 7.3 contains a list of over 200 existing visualization techniques. We have only listed a small subset, as there are hundreds of techniques from which to choose. As it is difficult to keep track of all possible techniques, this table can be used as a reference point during design. To find different types of visualization techniques, the interested reader can consult numerous sources from different disciplines: Aigner et al., 2011; Börner, 2010, 2015; Card et al., 1999; Chi, 2000; Coopmans et al., 2014; Cuoco and Curcio, 2001; Eilam and Gilbert, 2014; Few, 2013; Jonassen et al., 1993; Glasgow et al., 1995; Harris, 1999; Hansen and Johnson, 2005; Heer et al., 2010; Hentschel, 2002; Krum, 2013; Lankow et al., 2012; Lima, 2011, 2014; MacEachren, 1995; Malcolm, 2004; Markman, 1999; Meirelles, 2013; Moktefi and Shin, 2013; Pauwels, 2006; Peterson, 1996; Smiciklas, 2012; Spence, 2007; Telea, 2015; Tufte, 1983, 1990; Ward et al., 2015; Yau, 2012.

Table 7.3: A reference table for Stage 3 of the visualization design process: some visualization techniques

Visualization Techniques					
Activity bar chart	Control structure diagram	Gantt chart	Natal chart	Radar graph	Starburst plot
Adjacency diagram	Correlation graph	Gap chart	Nautical chart	Ragone chart	State diagram
Algebraic Petri net	Correlogram	Genogram	Network diagram	Railroad diagrams	Statistical map
Allele chart	Cremona diagram	Greninger chart	Node-link diagram	Ray map	Stemplot
Arc diagram	Chronological scale	Grotrian diagram	Nomogram	Recurrence plot	Streamgraph

Area cartogram	Cross-plot	Grouped surface graph	NP-chart	Recursive transition network	Strip map
Area chart	Dasymetric map	Half-bidirectional table	N-squared diagram	Reference table	Structure chart
Arrhenius plot	Data flow diagram	Heatmap	Nyquist plot	Run diagram	Sunburst chart
Arrow diagram	Dendogram	Hess diagram	Ogive graph	Run-sequence plot	Surface graph
Aviation map	Depth chart	Hexany diagram	Onion diagram	Sankey diagram	Symbol map
Axis scatter graph	Diagrammatic map	Histogram	Ontology chart	Scatter graph matrix	Syntax diagrams
Bachman diagram	Distorted map	Horizon graphs	Organigraph	Scatterplot	Tag/word cloud
Bar chart/graph	Dot distribution map	Hyperbolic tree	Orr diagram	Schematic diagram	Thematic map
Bathymetric chart	Dot plot	Hypsometric map	Overlay graph	Schulte table	Themescape
Bidirectional table	Eco-map	Index chart	Pantograph	Seasonal subseries plots	Time distance diagram
Binary decision diagram	Enclosure diagram	Influence diagram	Parallel coordinate plot	Self-similarity matrix	Tonti diagram
Biplot	Entitative graph	InfoCrystal	Pareto diagram	Sentence diagram	Topic map
Blobbogram	Entity–relationship diagram	Isarithmic maps	Parse tree	Sequence chart	Topological map
Block diagram	Euler diagram	Ishikawa diagram	Patch graph	Service planning diagram	Trace diagram
Bode plot	Existential graph	Isodemo-graphic map	Patch map	Silhouette graph	Trajectory line graph

Box plot	Explanatory map	Isoline graph	Pedigree chart	Similarity matrix	Transit map
Bubble chart	Exploded diagram	Isopleth map	Periodic tables	Simplex plot	Tree diagram
B+tree	Eye diagram	Jump line graph	PERT chart	Slice graph	Trilinear graph
Candlestick chart	Fan chart/ diagram	Junction table	Phase diagram	Skewplot	van Kampen diagram
Carroll diagram	Fishbone diagram	Karnaugh diagram	Pie chart	Small multiples	Venn diagram
Cartogram	Fishnet graph	Kernel density plot	Pin map	Snellen chart	Violin plot
Chess diagram	Flow map	Kohonen map	Piper diagram	Sociogram	Volume graph
Chord diagram	Flowchart	Line chart/ graph	Polar area diagram	Sparkline	Voronoi diagram
Choropleth	Forest plot	Linguistic map	Population pyramid	Spider diagram	Waterfall chart
Circuit diagram	Four-fold chart	Logical graph	Probability plot	Spie chart	Web chart
Cladogram	Fractal maps	Marked graph	Process flow diagram	Spin network	Wind rose graph
Commutative diagram	Frequency polygon	Mesh networks	Prognostic chart	Stacked graph	Zone diagram
Compressor map	Funnel chart	Mosaic plot	Proportional symbol map	Stacked histogram	
Concept map	Fuzzygram	Motion diagram	Qualitative map	Stacked pie chart	
Contingency table	Galbraith plot	Multiway table	Q–Q plot	Star plot/ graph	

7.1.4 STAGE 4: CONCRETE ENCODING AND INTERACTION

This stage is concerned with concrete encoding of the information items before a final visualization is settled on. It involves selecting concrete structures and marks, composing them together, choosing and applying visual variables, and bringing all of these elements together into a physical spatial system. At this stage, the designer has to determine how an adopted visualization technique should be implemented: what specific visual structures and marks should be used, and how these should be assembled together into a whole. Recall that a visual structure can be large and complex, being composed of many sub-structures, or can be simple and have few sub-structures. After creating these concrete structures, the designer needs to think about how to add more detail to these by selecting visual variables that will support the users' low-level visual tasks. For instance, assume that in Stage 3 the designer selected coordinate and multi-dimensional glyph structures, and decided on the scatterplot technique. In Stage 4, the designer would have to select a specfic coordinate structure (e.g., a 2d Cartesian coordinate system that has nominal scaling), a specific multi-dimensional glyph (e.g., a 2d starplot glyph), and specific variables (e.g., three different sizes—small, medium, large—and two color hues—red and blue). We wish to emphasize again that the choice of these structures, marks, and variables is dependent on the analysis done in Stage 1—i.e., the characteristics of the information space and the tasks and activities of the users. If the designer chooses a 2d Cartesian coordinate system with nominal scaling, it has to be consistent with the results of the analysis done in Stage 1.

We have already presented many types of visual structures, marks, and variables that can have specific instances—see Chapters 3, 4, and 5. Table 7.4 contains a list of possibilities for each of these elements of design. The interested reader can also consult the following sources: Bertin, 1967/1983; Börner, 2015; Cairo, 2012; Card et al., 1999; Carpendale, 2003; Cleveland and McGill, 1984, 1986; Few, 2009, 2013; Mackinlay, 1986; Mazza, 2009; Munzner, 2015; Nowell, 1997; Ware, 2008, 2012.

At this stage, if no interaction is needed, then the process can either end here, with the final product being a static visualization, or can iterate through the process again. Once Stage 4 is completed, the designer has a concrete visualization that can be implemented on different displays and platforms. Alternatively, for interactive visualizations, the designer needs to select low-level epistemic interactions that amplify and extend the semantic and communicative utility of the static visualization to further support the tasks and activities that were analyzed in Stage 1. In addition, the designer needs to analyze mirco- and macro-level aspects of interactivity (e.g., structural elements of each interaction). Table 7.4 contains a list of 32 possible epistemic interactions, a list of the micro-level structural elements of interaction, and macro-level interactivity factors from which the designer can choose, or which the designer should consider, in operationalizing the design of interaction (these were previously discussed in more detail in Chapter 6). For instance, one of the higher-level tasks may require users to reason about the temporal and spatial evolution of a stream of data, in which case the designer may consider the use of the *scoping* interaction. Furthermore, if the data has a very high sampling rate, the *flow* element of the scoping interaction may be opertionalized in a *continuous* fashion. As was suggested in Chapter 6, these interaction and interactivity constructs are still conceptual and need to be translated into low-level physical events. For a more in-depth analysis of interaction design for visualizations, the reader should consult the following sources: Elmqvist et al., 2011; Green and Fisher, 2011; Heer and Shneiderman, 2012; Pike et al., 2009; Roth, 2013; Sedig et al., 2012, 2013; Sedig and Parsons, 2013; Parsons and Sedig, 2013b; Tominski, 2015; Yi et al., 2007.

Table 7.4: A reference table for Stage 4 of the visualization design process: concrete encoding and interaction

Concrete Encoding and Interaction					
Visual Marks	Visual Variables	Interaction Epistemic Actions		Interactivity Micro Level	Macro Level
Dots	Size	Annotating	Accelerating/ Decelerating	*Action:*	Diversity
Dingbats	Shape	Arranging	Animating/ Freezing	Agency	Complemen- tarity
Letters	Texture	Assigning	Collapsing/ Expanding	Flow	Fitness
Digits	Hue	Blending	Composing/ Decomposing	Focus	Flexibility
Lines	Motion	Cloning	Gathering/ Discarding	Granularity	Genre
Circles	Luminance	Comparing	Inserting/ Removing	Presence	
Polygons	Saturation	Drilling	Linking/ Unlinking	Timing	
Arrows	Orientation	Filtering	Storing/ Retrieving	*Reaction:*	
...		Measuring Navigating Scoping Searching Selecting Transforming Translating ...		Activation Context Flow Transition Spread State	

7.2   VISUALIZATION DESIGN FOR COMPLEX COGNITIVE ACTIVITIES

As we stated before, even though the presented framework in this book can support design thinking to create static visualizations, its main utility is to help the designer create a host of visualizations that work in concert in the context of human-information interaction for performing complex cognitive activities, discussed in Chapter 6. To perform such activities, interaction with one visualization may result in the appearance of another visualization in the representation space; interaction with the latter visualization may result in the appearance of another visualization that encodes other aspects of an information space; and so on. For these activities, the designer should be able to generate not just one visualization, but many visualizations that support the needs of the users. Furthermore, the visualizations should be made interactive, and interactions should be coordinated appropriately. It is in such a context that we envision our framework to be most effective. Figure 7.2 is an adaptation of Figure 6.3, in which the role of the designer is added. The figure depicts the function of the design process in generating multiple interactive visualizations, and in supporting the cognitive activities of the user.

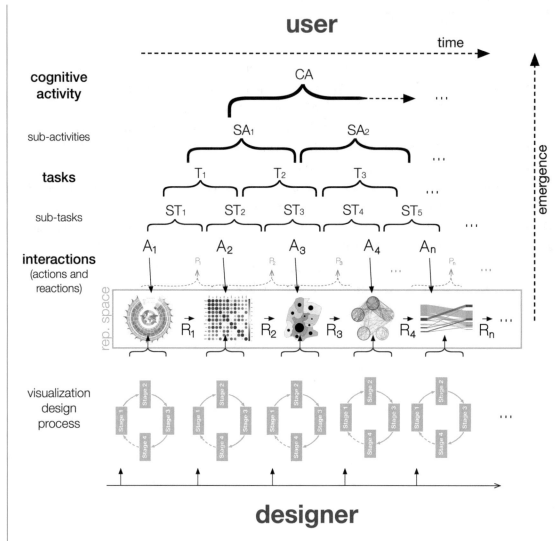

Figure 7.2: How the visualization design process can support complex cognitive activities.

CHAPTER 8

Application of Framework

The purpose of this chapter is to demonstrate the operational utility and application of our framework in the design of visualizations for human-information interaction. Here, we present three design examples, combining all the conceptual elements that have been discussed in the previous chapters. The examples take different angles in order to demonstrate different utilities of the framework. The first example illustrates the use of the framework in supporting systematic thinking about an information space, and how to map from information space to representation space using the design process presented in Chapter 7. The second example illustrates the use of the framework in the context of designing highly interactive visualizations. The third example highlights the utility of the framework when dealing with open-ended design situations in which the information space is large and complex, but the designer does not start with any pre-determined datasets. The three examples show how the framework can be used in the context of different domains of knowledge and activity: researching quality of life in different places in the U.S.; making diagnostic decisions using genetic profile of patients; and learning astronomy—galaxies, solar systems, planets, and so on. As it would be too lengthy to describe all aspects of the design process for each example, only certain parts of the process are described to highlight the utility of the framework. Additionally, the intention here is not to develop finalized visualizations; rather, it is to show how the framework can be used. Therefore, not all designs are finished and polished. This is done intentionally so as to keep attention focused on fundamentals and away from more surface-level details. Of course, in a real design situation, all of the details should be attended to appropriately.

In this chapter, when describing the design process, where appropriate, we will refer back to the stages described in Chapter 7. What we intend to convey here is the following: how the designer can think systematically about the information space in the context of systems; how the designer can think systematically about the representation space in the context of systems; how the designer can think systematically about how to map information items from an information space to a representation space; how the designer can proceed from high-level design, through the stages presented in Chapter 7, to low-level considerations, in a systematic fashion; and how the design process can also be thought about systematically with respect to interaction design and the intended users' tasks. Using the few examples in this chapter, our intention is to show the reader: how all 14 visualization patterns can appear in different designs; and how the proposed design process can help with systematic design thinking—from the most abstract, intention-driven level, proceeding down through lower levels of abstraction, and ending at low-level concerns related to the concrete appearance of visualizations in the representation space. By the end of this chapter, it should be clear that, even though the design process supports systematic thinking, it does not impose restrictions on designers, and allows for a great deal of flexibility and creativity in the final design.

8.1   QUALITY-OF-LIFE EXAMPLE

Consider a situation in which we want to design a set of visualizations to help research quality of life in different states in the U.S. Intended users could be interested in moving to a new state, or another county within their state, or could simply be interested in exploring and learning more about a state or a number of states. In the first stage of design, the designer needs to think about the information space. In this case, there is no specific dataset, database, or other source that is given to the designer. Rather, the designer has to gather information from a number of different places, and bring it together to comprise the information space. The designer can analyze this whole space from the perspective of systems theory, and use a spatial metaphor to conceptualize all of the relevant data and information. As discussed in detail in Chapter 3, there is considerable evidence that this type of thinking is useful in design. Hence, the designer needs to analyze the information space to think about what systems and sub-systems exist within it, what the key entities and relations are, and so on. For instance, within the space is geographical information regarding the U.S.—e.g., its shape, proportion, area, and boundaries. Each state, which can be considered as a sub-system of the country, has similar information. Moreover, each county, which can be considered as a sub-system of the state, has similar information. Thus, we can conceptualize these different features of the information space as systems, sub-systems, and so on. Within the information space also resides demographic data, climate data, education data, healthcare data, and so on. Figure 8.1 depicts the overall information space and some of its sub-spaces.

Following this, the designer needs to consider the goals of the users and the activities and tasks that they are likely to perform. Designers can think in a top-down fashion, considering the activities that users need to carry out to achieve their overall goals, the tasks that they would need to perform in order to carry out activities, the interactions that could be performed to achieve the tasks, and so on. For instance, in our current example, users' cognitive **activities** may include *exploring* and *learning* about states and counties, *investigating* the housing and crime situation in these states, and then *making a decision* about where to move; some **tasks** may be *comparing* crime rates in different states and counties, and *ranking* school districts; to achieve these, users can perform **interactions** such as *arranging*, *drilling*, and *filtering*; and, some of the necessary **visual tasks** may be *locating* counties within states and *identifying* desirable areas based on visual properties of representations. Of course, this type of analysis is done in light of the previous analysis of the information space.

Figure 8.1: Depiction of the overall information space and some of its sub-spaces.

The designer then begins to think about how to map information items from the information space to the representation space. Recall from Chapter 3 that "information item" refers to any entities, properties, relationships, systems, sub-systems, and so on, from the information space. What constitutes an information item in any given case depends on the designer's thinking and intention for information mapping—if she is thinking about the demographics sub-space as one individual thing, then it can be considered as an information item; if she is thinking about all of the sub-spaces of the demographics space separately, then each of them can be considered as an information item. In the former case, the designer may choose to map the item to the representation space by using the Token pattern, encoding the item in an individualized, unitized fashion; in the latter case, she may choose to map the items to the representation space by using a blending of any number of different patterns, such as the Hierarchy and Cell patterns, to encode the hierarchical nature of the items in a cellular, segmented fashion.

Let us assume that the designer wants to create a visualization that gives users an overview of eight sub-spaces that she deems important for users' activities and tasks—namely, those concerned with demographics, education, climate, housing, economy, energy, crime, and healthcare. The visualization should be composed of a number of sub-visualizations, where each encodes information items from the eight sub-spaces. As the designer intends the visualization to be highly interactive, she knows that her design has to be informed not only by higher-level activities and tasks, but also by lower-level epistemic interactions that users will want to perform on sub-visualizations in order to explore their underlying information (e.g., drilling, comparing, scoping, arranging, and translating). In addition to the eight sub-spaces, the designer also wants to map information pertaining to the geographic sub-space onto the representation space—information about different states and their respective counties—and allow users to select and compare a number of these items. Although at this stage the designer has a sense of the interactive possibilities of the visualization, interaction details can be left to later in the design process (i.e., Stage 4).

Once the information space and user tasks and activities are clear in the mind of the designer, at Stage 2 of the visualization design process—the abstraction space of design—the designer needs to think about patterns and structures and mapping-related issues in a conceptual way. She must think about which patterns to use, how they should be blended, and what abstract structures can support the pattern blendings. To do so, the designer may wish to systematically go through each pattern and think about design possibilities for different blendings. Alternatively, she may wish to think about specific items of the information space and their characteristics, and then determine which pattern(s) to use for mapping and encoding them. The designer has a great deal of flexibility in this regard, and may even wish to use other design frameworks or methodologies simultaneously.

In the current example, given the characteristics of the Cell pattern, the designer decides that she wants to use a cellular structure as the main overview in order to segment the sub-visualizations that will represent the eight sub-spaces. Each sub-visualization will itself be instantiated from a blending of different patterns. At this overview level, the designer also wants to represent a number of states and encode information about their shape and boundaries. The designer selects the Area pattern for this purpose. Additionally, the designer wants to show the partitioning of the state into counties and so selects the Cell pattern.

At Stage 3, the designer can select from existing representation techniques that instantiate the chosen patterns, or can create her own technique. Since the designer may not know of an existing technique that would instantiate her blendings, she does not worry about a label for her design. It is worth noting that the form at this level, as well as at the level of the final presentation in the representation space, is influenced by the designer's own aesthetic and stylistic preferences. Even at this stage of design, there are perhaps innumerable ways in which the patterns can be instantiated. For instance, the structure that instantiates the cells in the overview visualization can be any shape (e.g., square, circular, polygonal, or another shape). Let us assume that the designer has chosen to instantiate this Cell pattern as circular donut slices along with a circle. Figure 8.2a shows an example structure that is [CL•TK]-based, where the Token pattern is used for the names of the sub-spaces. Each cell acts as a container for the sub-visualizations that represent each sub-space and encode its information items. The cell in the center is intended to be used for placing central visualizations for analysis and comparison of states.

At this stage, the designer knows she also wants to use the Area and Cell patterns to represent the U.S., its states, and their counties. Let us assume that users will be presented with an [AR•-CL]-based visual representation (e.g., a geographic map) of the country and can choose a number of states in which they are interested. The designer wants these selected states to be shown along with the Cell-based structure—not as sub-visualizations but as sibling visualizations, as they can be considered as sibling sub-spaces within the information space. Figure 8.2b shows how this can be implemented alongside the existing structure, assuming a user has already selected six states. The outer visualizations can be thought of as metaphorically orbiting the inner visualization structure.

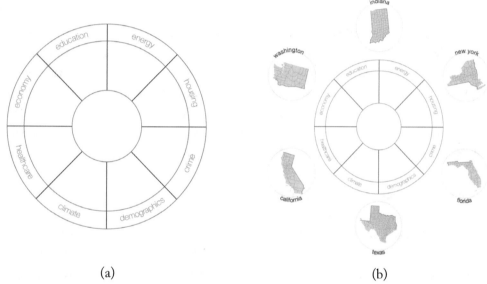

(a)   (b)

Figure 8.2: (a) Visual structure that is based on the Cell and Token patterns; (b) additional [AR•-CL]-based visualizations are placed around main structure along with TK-based labels.

Once the designer has settled on this overall visual structure, she then must decide on the sub-visualizations that will populate it.[10] For the sake of space, we describe in detail the design of only one such visualization, but the design process for the others is similar. The representation we will describe is for the healthcare sub-space. At this point in the design process, the designer may return to Stage 1 and analyze this sub-space. An important piece of information from this space, which the designer knows users want to examine, is healthcare visits and wait times, for both emergency rooms and referrals. To decide how to develop an appropriate visualization for this sub-space, the designer proceeds to Stage 2 to select and then blend more patterns. As she consults the patterns, she thinks about which ones are useful for the current context, and can brainstorm and

---

[10] The designer of course could have done it the other way around—first creating a set of visualizations, then determining a super-structure visualization in which to place them. We simply illustrate one possible design process.

reflect on them in light of the intended user tasks. For example, she wants to somehow represent healthcare visits with respect to time. She wants to map the visits onto visualizations such that they are each individualized and unitized—that is, she wants to make use of the Token pattern. She also decides to use the Coordinate pattern to give them an external frame of reference with respect to time. Additionally, she wants to represent different types of visits (e.g., ER or clinic visits). With respect to ER visits, she knows that the wait times are dependent on the severity of the case, and thus also wants to encode this aspect of the sub-space. She decides that she needs the Hierarchy pattern to show this parent-child type of relationship. Finally, in order to represent the channel- or path-like nature of the visits, she decides that the Track pattern is also needed. Thus, at this stage she has selected Token, Coordinate, Hierarchy, and Track as the underlying patterns for the desired visualization (i.e., a [TK•CR•HR•TR]-based representation). Figure 8.3 shows the generated visualization. The overall structure is based on the Track pattern, which is self-blended. Specific events, such as arrival and discharge, are encoded as simple dots. They are placed within four main tracks: arrival (A), waiting (W), visit with a doctor or nurse (V), and discharge (D). The waiting and visit tracks themselves have sub-tracks labeled 1–4 in order of increasing severity of the case. A Coordinate-based structure at the top gives a temporal frame of reference at an hourly granularity.

After generating this visualization, the designer realizes that the duration of time between specific events may be difficult to determine perceptually, even though they are in separate tracks. As mentioned in the previous section, designers can and are encouraged to iterate their designs, and to go back and rethink their choice of patterns and blendings. As the designer proceeds through Stages 1 and 2 again, she realizes that the Link pattern is also important here, as she wants to encode explicit connections between events within a track. Thus the visualization becomes [TK•CR•HR•TR•LK]-based. Figure 8.4 shows the result of instantiating the Link pattern in the visualization.

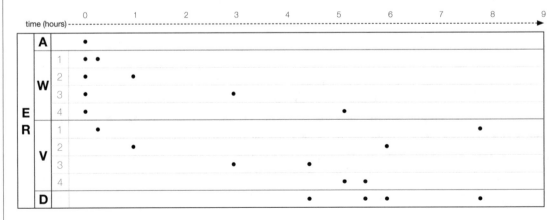

Figure 8.3: First iteration of the visualization for the healthcare sub-space showing waiting times for emergency room visits.

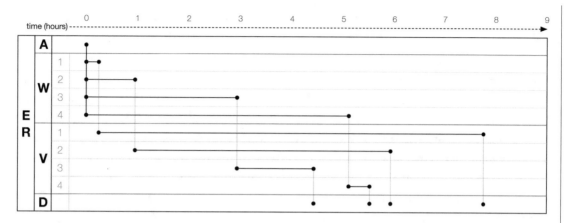

Figure 8.4: Second iteration of the visualization for the healthcare sub-space, with the addition of the Link pattern.

Now, back in Stage 4 again, the designer needs to consider the visual variables that are employed, and other low-level details regarding color, texture, layout, and so on—especially with respect to the visual tasks that users will likely perform. Although many decisions occur at this stage, we only mention a couple for the sake of brevity. One decision is with respect to the coloring of the visualization—of especial concern is coloring the sub-visualizations (i.e., visual marks) such that their distinctness is made clear with respect to the track to which they belong in the "W" and "V" tracks. This is assuming that users will want to perform visual tasks such as *following a path* as it crosses into other tracks. For example, the designer knows that the most distinct hues are yellow, green, red, and blue (see, e.g., Ware, 2008), and thus chooses to use them for optimal visual distinctness. Figure 8.5 shows the result of this design decision.

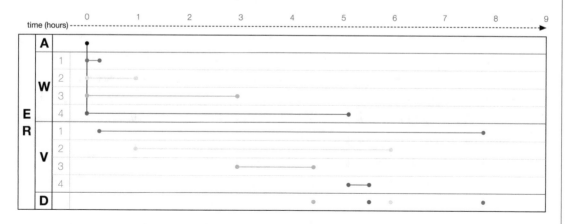

Figure 8.5: Adjusting surface-level details of the visualization (i.e., color) for optimal visual distinctness.

At this point, the designer again reconsiders the visual tasks that users are likely to perform in light of the chosen visual variables. For instance, if users want to visually reason about overall paths from arrival to discharge, the visualization in its current state is likely sufficient. However, if users want to visually compare times between the "W" and "V" tracks for the same severity level—e.g., compare the length of time in waiting and the length of time in visiting for level 1 severity—then the visualization in its current state is not optimal. As discussed in Chapter 3, comparing the length of lines with different baselines is not ideal if a high degree of accuracy is required. Figure 8.6 shows a visualization that better supports such a task. Anticipating that the aforementioned visual task is likely to be performed, the designer can provide interaction mechanisms to make this particular visualization available when users want it. The designer has chosen to use the Token, Link, Coordinate, and Group patterns for this visualization (the visualization is [TK•LK•CR•GR]-based). In viewing the visualization, users can perform accurate visual comparisons of the wait and visit times for the different severity levels. While there are other considerations at this level of the design, for the sake of brevity they are not described here, and are left as an exercise for the interested reader.

Figure 8.6: A special visualization that facilitates certain perceptual tasks and that can be displayed through interaction.

Once the designer has decided on this visualization, she can go back to the overview visualization (shown in Figure 8.2) for which this is to be a sub-visualization. She can engage in a similar process for the other seven sub-spaces of the overall information space—i.e., those related to demographics, education, climate, housing, economy, energy, and crime. For the sake of space, we will not go into the same level of detail about the design process for those sub-spaces. The purpose of this section is simply to illustrate how the design process helps with systematic design, and not to fully describe a design process. That being said, we will give very brief descriptions of patterns the designer may choose for two more of the sub-spaces.

- Demographics: This sub-space is itself composed of a number of sub-spaces with respect to language, ethnicity, age, income, religion, and so on. Let us assume the designer wants to map one of these sub-spaces onto visualizations and encode the change and variation of languages over time. To do so, the designer takes a number of information items (people and their spoken language) and maps them onto visualizations to make them integrated, continuous, and indistinguishable—individual people cannot be distinguished in the visualization—thus choosing the Fusion pattern. The designer also wants an external temporal frame of reference, and chooses to use the Coordinate pattern. In addition, the designer wants to encode the co-occurrent nature of the languages, and thus chooses the Stack pattern. As a result, the designer chooses a visualization that is [FS·ST·CR]-based. There are a number of existing techniques that are based on this blending, and the designer decides to instantiate the blending with a streamgraph technique.

- Education: This sub-space is composed of geographic information (e.g., states and counties) as well as rankings of the counties based on ratings of high schools within them. The designer wants to encode the variation and ranking of counties within the states that the user has chosen, so that the user can identify the location of good and bad counties with respect to school ratings. Additionally, the designer wants to map items (specific counties) onto visualizations and organize them in a sequential fashion to encode their ranking. As a result, the designer selects the Spectrum, List, and Token patterns—i.e., she wants to create a [TK·SP·LS]-based visualization. Figure 8.7 shows one way in which this can be blended with the existing patterns in the abstract superstructure-visualization. The user can interact with the visualization in the center to select what should be shown—in this case, the top five counties in each state.

Visualizations for the other sub-spaces can be created by following the same process as the previous ones. Figure 8.8 shows a possible design of the overall visualization after designing sub-visualizations for each of the sub-spaces. As can be seen, the overall visualization is based on a blending of all of the patterns presented in Chapter 4. For example, in addition to the previously described sub-visualizations, the energy sub-visualization is [BR·TK]-based, the housing visualization is [HR·CL·TK]-based, the climate sub-visualization is [CL·LK·TK]-based, the crime sub-visualization is [SP·AR]-based, and the economy sub-visualization is [ST·FS]-based.

It is important to note that if this were a static visualization, the sub-visualizations would have limited utility, as they are very small and lack detail. When the visualization is made interactive, however, the user can interact with each of the sub-representations individually—to increase their size, translate them into other visualizations, filter them, and so on. In each case, the different sub-visualizations can be temporally linked through interaction. It is for this reason that we have previously stated that the real power of this framework is in interactive contexts. Viewing the representation space as a system allows for precise thinking about which components (i.e., sub-visualizations) can and should be acted upon, and how other visualizations at the same or other levels should be affected by such actions (see Chapter 3 for more on this issue). Furthermore, as discussed in Chapter 3, because the designer thinks of visualizations as systems, it is easy to think about applying visual variables to sub-visualizations as well as to the overall visualization itself. For example, while we have previously talked about lower-level design considerations with respect to the health-care sub-visualization, the designer could apply similar design choices to the overall visualization. Figure 8.9 shows the outer circles and states adjusted such that their size encodes the population of the state. Similar design choices can be made with respect to color, shape, orientation, and so on.

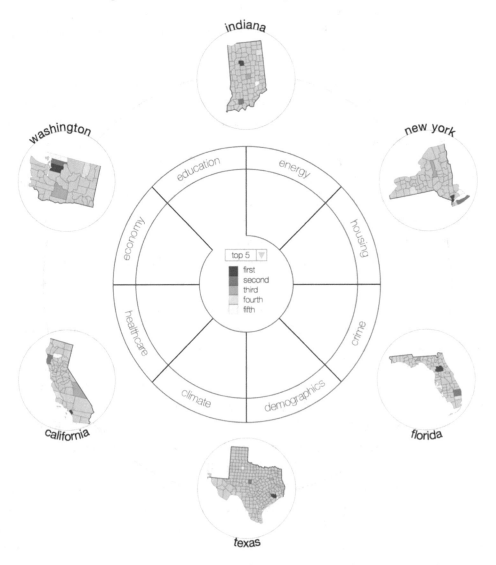

Figure 8.7: Instantiating the patterns for the education sub-space within the overall visualization.

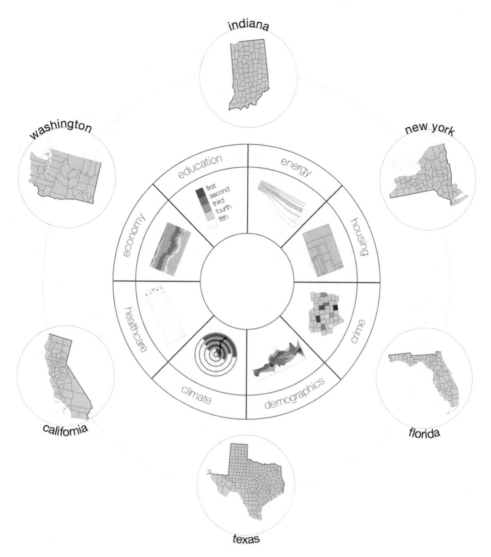

Figure 8.8: Overall visualization, which is based on all the patterns in Chapter 4. Can be interacted with to engage in deeper discourse with the data.

Figure 8.9: Encoding information items using visual variables of sub-visualizations; size of outer circles and states encodes population.

8.2   GENOMICS EXAMPLE

In the first example, we focused on illustrating how the framework enables systematic design. In this example, we focus on illustrating design decisions in the context of interactive possibilities. When designing static visualizations, designers have to think about encoding all relevant information items in one visualization. Such is the case when designing information graphics for magazines, for example. When designing interactive visualizations, however, designers can think about encoding the information items in many different visualizations, each of which can appear, disappear, transform, and/or be interacted with in a number of different ways.

In presenting this example, we do not discuss in detail the thinking behind selecting and blending patterns. We simply mention the use of some patterns, techniques, visual marks, and so on, with the assumption that the designer has gone through the relevant stages of the design process. We focus instead on design thinking with respect to visualizations for human-information interaction (i.e., conceptualizing information items within an information space, mapping them to a representation space, and designing interaction with the representations). Thus, we focus mostly on Stages 1 and 4 of the design process presented previously, with minimal discussion of the stages in between them. As was the case in the first example, we do not intend to show completely finished, polished visualizations that have all details accounted for.

Let us assume the designer is generating a set of interactive visualizations for a genomics application. The information space is large and complex and includes general genomic information (e.g., genes, phenotypes, disorders, pathways, and chromosomes) as well as specific information about a set of people (e.g., genetic variations, phenotypes, and inheritance). Intended users are clinical geneticists and cytogeneticists whose overall activity is to make diagnostic decisions based on genetic information of patients. The set of people in the information space are patients that have been referred to a geneticist. User tasks include:

- identifying common features (i.e., phenotypes) among individuals;

- determining if individuals have a known syndrome or disease;

- determining the exact location of genetic variants (e.g., copy number variants) of the individuals;

- assessing the relevance of genetic variants to known diseases; and

- identifying genes that exist within the region of the variants.

The designer plans to provide a number of different visualizations to support the above tasks. Moreover, the designer knows that some of the tasks depend on the completion of others, and thus plans for the visualizations to unfold in a sequential fashion as the user interacts with them. As the designer conceptualizes the structure of the information space, she also thinks about how to map information items to the representation space, and, in particular, how much of the information to encode at each stage. Recall from Chapter 3 that a simple visual mark can represent a large, complex sub-space, while encoding only very minimal information. For example, the information space under consideration contains sub-spaces pertaining to each individual. Each of those spaces can be considered as a singular information item, and the designer can encode the existence of that item in a simple dot. Each dot represents an individual and all of his or her attendant information, but encodes very little of it—only the *existence* of the individual. Figure 8.10 demonstrates this. On the left is a depiction of the designer's conceptualization of the information space, where each "*" represents an individual. As the designer is thinking about each individual as a unitized, singular entity—rather than as a complex space of properties and relationships pertaining to the individual—she wants to use the Token pattern and instantiate it with a simple visual mark (i.e., a dot). The right side of Figure 8.10 shows the result of the design process, where the designer has blended and instantiated the Cell and Token patterns using the Venn diagram technique. Each circle in the visual structure encodes a specific phenotype (P1, P2, or P3); each dot encodes the existence of an individual; and the containment of dots within the circles encodes the presence of the phenotype for the individual.

Because the designer knows about the complex sub-space underlying each visual mark, she can have its information items encoded and "brought to the surface" of the representation space through interaction. Additionally, because the designer has some domain knowledge and understands how users perform their tasks, she knows that this visualization will be useful as a starting point for users.

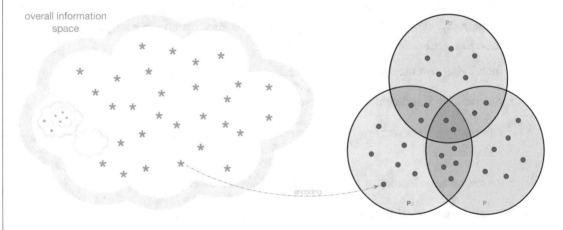

Figure 8.10: Mapping information items from the information space (L) to the representation space (R).

Once users view the visualization in Figure 8.10, they will likely want to drill into one of the individuals to access his or her information sub-space (e.g., genetic information). Thus the designer creates another visualization to encode relevant information items from the sub-space. Figure 8.11 shows this visualization and its underlying sub-space. The left side depicts how a single information item from the overall information space is now conceptualized as its own sub-space. By doing so, the designer thinks about its attendant information items, and how to map them to the representation space to suit other user tasks. On the right side, the designer has created a

[CL•CR•TK]-based visualization that encodes genetic variants of the individual, the chromosome on which they reside, interval length and size, number of genes within the interval, and known inheritance information.[11]

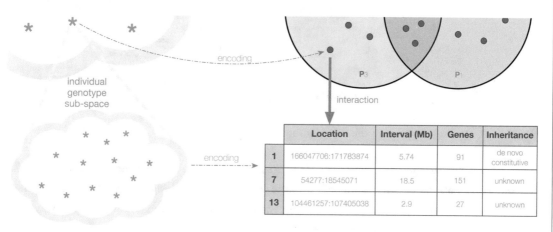

Figure 8.11: A user can drill into a visualization to open it up and reveal its underlying data.

At this point in a user's activity, the designer knows he or she is likely to want to identify genes that exist within the chromosomal location of the variant in order to reason about possible genetic connections. Thus, the designer provides interaction mechanisms to select and drill into one of the displayed locations. Figure 8.12 shows the result of this interaction. In this stage of the activity, information items from a different sub-space—the sub-space related to general genomic information—are being encoded. The designer has generated a visualization that is [TK•CR•TR•ST]-based, which depicts the relevant chromosome, its specific interval, and some features that exist within the interval. Each feature (e.g., gene, CNV) is instantiated in a Track-based structure, which is typical of genomic visualizations (see the example in Chapter 5).

---

[11] The figures in this section are not meant to depict the exact look of the finished representation space. For instance, the tabular visualization in Figure 8.11 does not necessarily appear directly underneath the Venn diagram in the actual finished design. The point is simply to depict the relationships between the different visualizations, and how they are brought about through interaction.

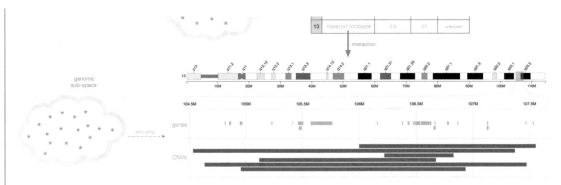

Figure 8.12: Drilling deeper into the information space.

At this juncture in the activity, the user can go back "up" to interact further with any of the previous visualizations, or can drill further into the current one. The designer knows that sometimes, when the user is thinking about and interacting with genes, information about gene pathways can be useful. Thus, the designer decides to encode the gene pathway sub-space in latent, potential form for each gene. Users can interact with the gene visualization to access the pathway visualization. Figure 8.13 shows the result of such an action. The designer has encoded the items in the pathway sub-space by mapping them onto a structure that is [LK•TK]-based—a directed tree. The designer has used some visual variables (e.g., color and shape) of the Token-based visualizations to encode information about the types of entities in the pathway sub-space (e.g., gene, protein, and molecule).

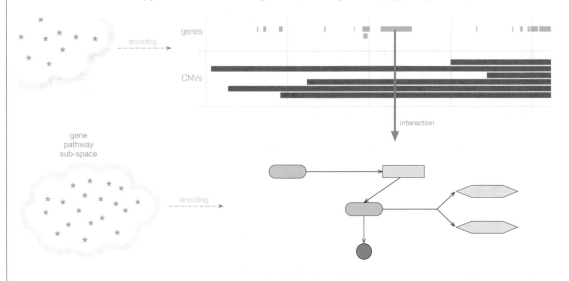

Figure 8.13: Drilling deeper to encode gene pathway information.

Conceptually, the designer can think of the visualizations as being latent and embedded within one another, progressively unfolding through interaction by the user. Alternatively, the visualizations can be thought of as underlying one another and being brought to the surface of the representation space through interaction. Essentially, designers can think of visualization design in a multi-faceted, embedded, and multi-level fashion. Combined with frameworks for interaction design, designers can engage in a systematic, holistic design process for human-information interaction. Figure 8.14 depicts the multi-leveled approach to the design of a genomic information space and some of its sub-spaces, and shows how user interaction can reveal layers of latent detail by accessing embedded visualizations.

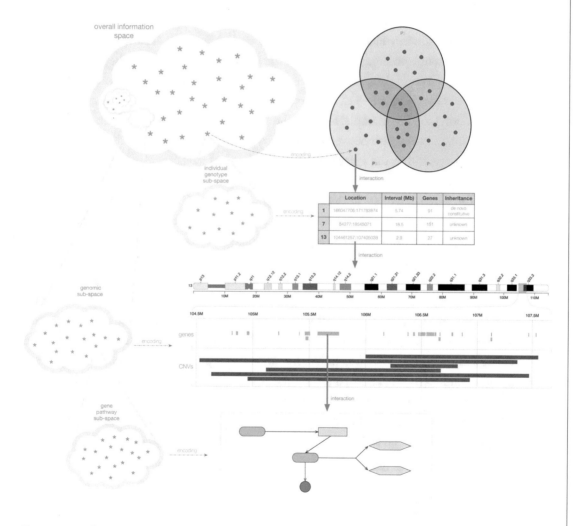

Figure 8.14: Comprehensive view of a user drilling deep into the information space through interaction. The designer can use different design patterns at different stages of the human-information discourse.

8.3   EDUCATION EXAMPLE

In the first example, our focus was on illustrating how the framework supports systematic thinking about an information space, and how to map from an information space to a representation space using the design process presented in Chapter 7. In the second example, we focused on how the framework can help with design decisions in the context of interactive possibilities. Here we present an example from the education domain. We aim to illustrate how the framework can help with open-ended design situations, where the information space is large and complex, and the designer has to conceptualize the information space and the users' tasks before determining what data can and should be used, and whether visualizations should be data-driven or model-driven (see Section 3.3 for more on this distinction).

Consider a situation in which a designer is hired to develop a visualization tool that should help students learn about astronomy. In this case, the design activity is not limited to generating visualizations for a given set of data, as no specific datasets are provided to the designer. Rather, a set of learning objectives are provided—i.e., descriptions of what students should know after working with the visualization tool. For example, some of the learning objectives may be: identify the planets in the solar system; distinguish between different types of planets based on their composition; locate the solar system within the galaxy; identify the types of galaxies; identify the type of the galaxy in which we live; rank the planets according to size and distance from the sun; describe the composition of the different planets in the solar system; and so on. Given these tasks, the designer needs to conceptualize the information space, gather relevant data, decide what sub-spaces should be model-based, determine what should be represented, plan for interaction possibilities, and so on. In such scenarios, the design activity is open-ended and loosely defined, and a designer can benefit from a framework that helps to guide his/her thinking about the activity and its constituent tasks.

In Stage 1 of the design process, the designer analyzes the information space and the task space. The information space can be conceptualized as a complex multi-level system. For example, a galaxy can be viewed as a system, with entities (e.g., solar systems), relations (e.g., distances), and properties (e.g., size); a single solar system is a sub-system of a galaxy, being composed of planets, a star, and other entities and relations; a single planet is a sub-system of the solar system, being composed of different entities (core, mantle); and so on. Thus the information space can be viewed as a system composed of many levels. To improve the analysis, the designer can ask a number of questions about the information space, such as those listed in Section 7.1. For example:

- What are the types of information items (e.g., how can planets be classified)?

- Are there existing datasets that can be used? If so, are they structured or unstructured? Homogenous or heterogeneous? Large or small?

- Is the information space static? Or will there be derived information added to the space, or new information coming into the space (e.g., data streams)?

- Are there information items that are not data-driven (i.e., real data obtained from galaxies), but rather based on existing generic models of galaxies?

Conceptualizing the information space should occur along with an analysis of the users' activities and tasks. In this case, tasks will be performed to achieve the goals of various learning objectives. As the designer thinks about the different tasks that may need to be performed, he can think about how the information space should be conceptualized—e.g., what should be considered an information item? What level of granularity is useful for the intended users? The designer can also ask questions about the sub-activities that need to be performed—e.g., will users need to explore, predict, make decisions, solve problems, triage, and so on?

In Stage 2 of the design process, the designer has to decide which patterns to select and blend. There are a few ways this can be done—for example, one way is to choose a task or objective, and go through the patterns to brainstorm that would be useful in supporting the task. The designer can choose a task, such as being able to rank the planets in our solar system according to size, distance from the sun, and length of orbit, and identify which patterns would help in this regard. Another way is for designers to go through the list of patterns, examining each one's utility, to see if it might be useful in relation to the learning objectives and then choose them accordingly.

The designer wants users to be able to learn about the hierarchical, embedded nature of the universe, and the Earth's place within the universe at different levels of granularity—e.g., within the solar system, within the galaxy, within a galaxy cluster, and so on. Hence, the designer chooses the hierarchy pattern. Given this pattern, the designer creates a visual structure in which the sub-visualizations are circular coordinate structures ($\in$ CR) that contain analogical token-based representations of individual galaxies ($\in$ TK). These token-based representations are mapped in such a way that their region and spatial dimensions are encoded ($\in$ AR). Hence, this portion of the visualization is [TK·CR·AR·HR]-based. The designer also knows that one of the learning objectives is for users to be familiar with some of the better-known galaxies, and the galactic groups to which they belong. This goal fits well with the previous one, and the designer decides to place another visualization alongside the first one. To achieve this goal, the designer decides to map information items (galaxies) onto the visualization by placing their labels ($\in$ TK) within a cell-based structure, thus emphasizing their categorization by galactic group, hence obtaining a [CL·TK]-based representation. Figure 8.15 shows these two representations, which together form a [TK·CL·CR·AR·HR]-based representation system.

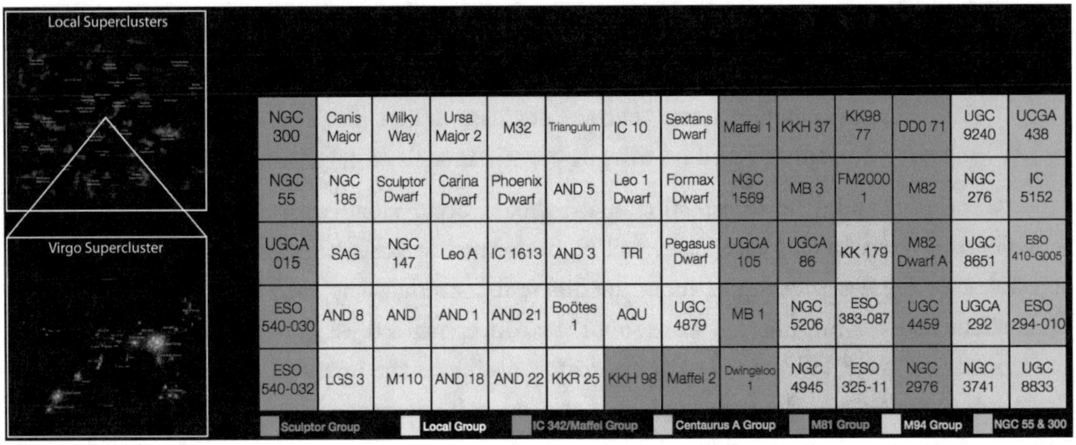

Figure 8.15: Two visualizations that together form a [TK•CL•CR•AR•HR]-based visual representation. Galaxy clusters courtesy of Andrew Z. Colvin, commons.wikimedia.org/wiki/File:Earth%27s_Location_in_the_Universe_(JPEG).jpg.

As the two visualizations are linked, the user can interact with the one on the right to drill deeper into and unfold the visualization on the left. These two visualizations can be conceptualized as two sub-systems of a higher-level system. In Figure 8.16, below, the user has acted upon the sub-visualization on the right by selecting the Local Group. This has caused the Local Group to appear in the sub-visualization on the left, helping the user see the location of the group within the supercluster and understand further the embedded, hierarchical nature of the entities under investigation.

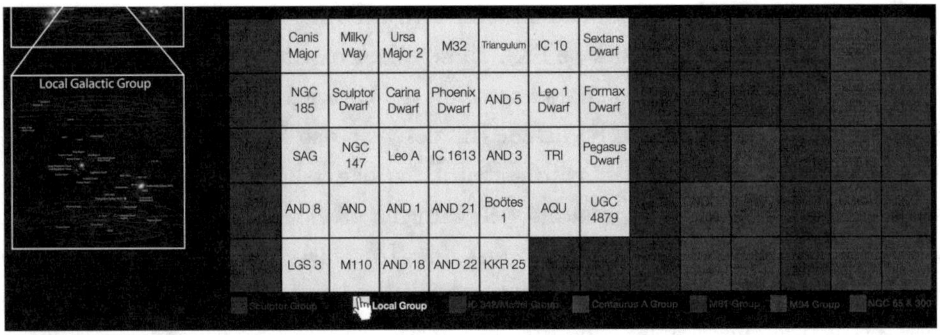

Figure 8.16: The user has selected the Local Group, causing it to become encoded on the left. Galaxy clusters courtesy of Andrew Z. Colvin, commons.wikimedia.org/wiki/File:Earth%27s_Location_in_the_Universe_(JPEG).jpg.

As the intention is to design an interactive visualization, the designer considers how epistemic interactions can be combined with the patterns to fulfill the design goals (i.e., helping students achieve the learning objectives). Figure 8.17 shows the result of the user acting upon three of the cells to bring them into focus and move the others into the periphery, while bringing new information into the representation space. This interaction technique, sometimes referred to as semantic zooming, is usually a combination of two epistemic actions: drilling and transforming. The designer has provided the user with the ability to act upon the visualization to bring more information to the surface—i.e., the image of the galaxy—as well as the ability to manipulate the semantic and geometric properties of the visualization. By increasing the size of the selected cells, and decreasing the size of the others, the overall size of the visualization remains constant, while the cells in which the user is interested increase in size to allow more information to be represented. Based on the learning objectives with which the designer is working, he has decided to use an analogical representation with a high degree of fidelity (i.e., an image) of the galaxies to help students understand what they look like. The three selected galaxies are also highlighted in the visualization on the left, helping the user identify their locations within the group.

Figure 8.17: The user has drilled into three cells and transformed the representation space. Galaxy clusters courtesy of Andrew Z. Colvin, commons.wikimedia.org/wiki/File:Earth%27s_Location_in_the_Universe_(JPEG).jpg.

The designer knows that one of the objectives is for students to be able to identify and describe characteristics of different types of galaxies, as well as their growth and evolution. They should be able to develop an understanding of the evolution of galaxies from dense clusters of stars to identifiable shapes—e.g., spirals and barred spirals. To support this, the designer maps information items (generic stars) onto simple token-based representations (small circles), and places them in dense clusters that represent model-driven generic galaxies. The designer intends for the behavior of the circles over time to help the user learn about the evolution of galaxies. After exploring different possibilities for representing their evolution over time, including an examination of epistemic action patterns, the designer decides to use the scoping interaction, which enables users to interactively adjust the growth and development of the galaxies. In order to maintain context in the user's reasoning activities, the designer chooses to couple this visualization with the [CL•TK]-based visualization described above. Figure 8.18 shows model-driven generic galaxy types below each selected galaxy—barred spiral, elliptical, and spiral, for Milky Way, M32, and Triangulum, respectively. As a result of this design choice, the user can view a realistic image of the galaxy while simultaneously scoping the evolution of a model galaxy of the same type.

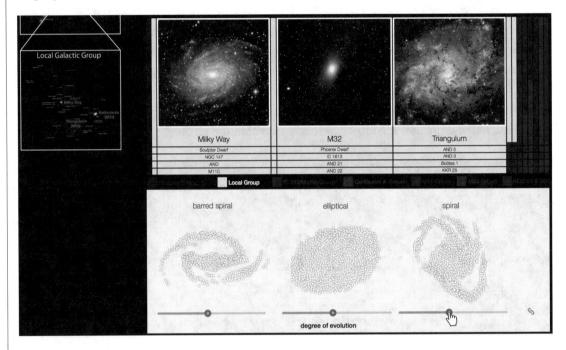

Figure 8.18: Exploring the evolution of different types of galaxies: barred spiral, elliptical, and spiral, for Milky Way, M32, and Triangulum, respectively. Galaxy clusters courtesy of Andrew Z. Colvin, commons.wikimedia.org/wiki/File:Earth%27s_Location_in_the_Universe_(JPEG).jpg.

From this point, the user can act on the visualization on the left to get more of the underlying, latent information to be encoded, thus showing the hierarchical nature of systems in the universe at multiple levels. The solar system is one of the main areas of focus for students (i.e., main learning objectives)—e.g., its place in the interstellar neighborhood and in the Milky Way Galaxy; its planets and their characteristics; and so on. Figure 8.19 shows this stage of the visualization, where this hierarchical system is now shown on the left and where the user has opened up the solar system to see the planets and the main moons. To facilitate the objective of being able to identify the ordering of the planets, the designer has decided to use the list pattern, where the planets are mapped analogically to small token-based representations, organizing them sequentially according to their distance from the sun—hence a [TK•LS]-based representation. Students are also supposed to be able to identify the major moons of the solar system, and understand some of their properties and relationships. To facilitate this, the designer has chosen to blend the coordinate and token patterns to provide a frame of reference for the moons and distances from their respective planets. Thus, the solar system part of the visualization is [CR•TK•LS]-based. At this stage in the user's activity, we can see how the representation on the left provides a hierarchical, embedded view of the solar system within the universe. This is an objective view, depicting our place within the universe. Along with this, it is important for students to learn another view of the universe, and be able to identify stars, constellations, and other entities, as they are seen from earth. The designer decides to provide this in one complex, yet coherent, visualization. In Figure 8.19, the upper right portion contains an interactive representation of space as it is seen from the earth. The representation is based on a blending of the coordinate, link, token, area, and spectrum patterns and is [CR•LK•T-K•AR•SP]-based.

Students can interact with the representation in the top right portion of Figure 8.19 to further support their learning activities. Figure 8.20 shows the result of the user selecting a region of the space, causing the region to enlarge and the representation to shrink and function as a context view. The user can then learn about specific constellations (e.g., their size and shape, constituent stars, and so on).

As the overall visualization is highly complex, encoding a great deal of information, the designer knows that providing a number of different interaction possibilities is critical to ensuring successful learning outcomes. Each sub-visualization has its own appropriate interactions incorporated into it. For example, to learn about the inner composition of the planets, the user can act upon a planet to drill into it and bring more information to the surface. Figure 8.21 (L) shows a cross-section of the different components of the earth. Figure 8.21 (R) shows the result of the user translating the sub-visualization into a conceptually equivalent form, in which the same components are shown, but without the detail of the earth's outer surface, and with some added measurement detail.

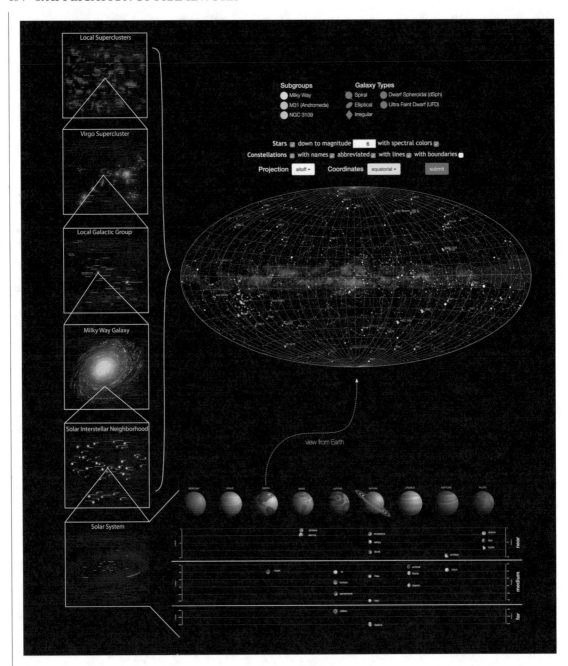

Figure 8.19: Large, complex visualization that has unfolded through interaction. The visualization is based on a blending of many patterns. Planets designed by Freepik.com. Interactive celestial map courtesy of github.com/ofrohn/d3-celestial. Galaxy clusters courtesy of Andrew Z. Colvin, commons. wikimedia.org/wiki/File:Earth%27s_Location_in_the_Universe_(JPEG).jpg.

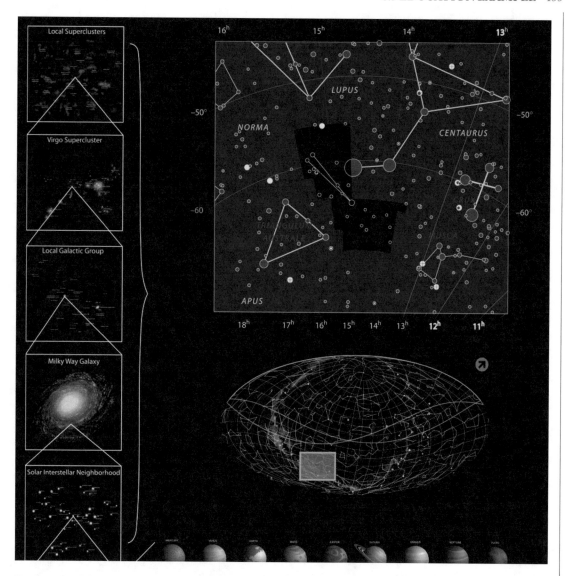

Figure 8.20: The user has selected a region of space in order to explore a set of constellations in more detail. Constellation graphics courtesy of IAU and Sky and Telescope magazine (Roger Sinnott and Rick Fienberg). Planets designed by Freepik.com. Interactive celestial map courtesy of github.com/ofrohn/d3-celestial. Galaxy clusters courtesy of Andrew Z. Colvin, commons.wikimedia.org/wiki/File:Earth%27s_Location_in_the_Universe_(JPEG).jpg.

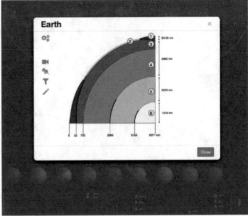

Figure 8.21: The user drilling into the Earth to view its inner composition (L). The user can act upon the visualization to translate it into a conceptually equivalent form (R). L courtesy of Kelvinsong, commons.wikimedia.org/wiki/File:Earth_poster.svg. R courtesy of Dake, commons.wikimedia.org/wiki/File:Slice_earth.svg.

CHAPTER   9

Discussion and Summary

In recent years, visualization design has received increasing attention from researchers and practitioners in a number of different fields, including statistics, computer science, cartography, business intelligence, information science, library science, journalism, medical and health informatics, and education. Research conducted throughout the past couple of decades has led to the development of many principles, guidelines, and heuristics for visualization design. Additionally, many techniques—such as treemaps, streamgraphs, heatmaps, and parallel coordinate plots—have been developed. Such techniques complement the set of long-standing and familiar ones such as scatterplots, line and bar graphs, and Venn diagrams.

Although previous work has attempted to classify visualizations in a number of different ways, it remains difficult to capture and characterize all possible visualizations in a manner that is useful for supporting systematic design. This is partly due to the fact that there are thousands of existing visualization techniques, and new ones are continually being devised to deal with new domains, users, information spaces, data types, and technologies. Moreover, although helpful for organizing and making sense of the existing design space, current classifications often do not help designers generate visualizations systematically in novel and creative ways. Rather, designers often implement known, existing techniques that are popular, familiar, or recent. Research in design studies suggests that this is especially true of novice designers. While experienced designers will begin a design process with abstract conceptualizations, novices tend to gravitate toward concrete solutions from the very start. Cognitive biases, such as the availability heuristic, can also work against design creativity by calling to mind familiar and recently seen techniques, thus making it difficult to start design from a "blank slate," and not be constrained by prior conceptions and known techniques. Conceptual frameworks can help scaffold design by organizing a space of ideas and supporting systematic design thinking.

In this book we have presented a framework for visualization design for human-information interaction, composed of a number of conceptual elements that are intended to aid in design thinking: *systems theory* and the attendant concepts of system, sub-system, super-system, properties, relationships, and levels; *spaces* and the attendant concepts of information space, representation space, sub-spaces, mapping, and information items; *levels of abstraction* and the attendant concepts of visualization techniques, structures, marks, and variables; a *pattern language* and its attendant concepts of patterns, blending, nesting, abstract structures, and instantiation; *human-information interaction* and the attendant concepts of joint cognitive systems, activities, tasks, interactions, events, epistemic action patterns, and interactivity elements and factors; and a *design process*, which integrates the aforementioned conceptual elements in the context of multiple design stages and design iteration.

The utility of these conceptual elements can be briefly summarized as follows. As a conceptual lens, systems theory supports designers in viewing design situations with a consistent set of concepts and terms for characterizing and articulating different features of the design space. The language of systems theory allows for consistent characterization at a general level, affording designers a common vocabulary for discussing aspects of visualization design. The idea of different spaces can promote spatial thinking and the use of spatial metaphors, which are known to be beneficial in design. Moreover, thinking about visualization design as mapping from one space to another—information space to representation space—can support design thinking, particularly when combined with the concept of patterns and when using the systems lens. Engaging in the visualization design process by thinking in terms of multiple levels of abstraction—gradually moving from highly abstract to more concrete—can help designers structure their thinking in a clear manner. Experienced designers consistently demonstrate the ability to traverse multiple levels of abstraction during a design activity. The core of the framework is our pattern language. The patterns that we have identified are basic and abstract, and function as conceptual tools that designers can use to aid in thinking about mapping information from an information space to a representation space. The relatively small number (14) of patterns is manageable to deal with and, when used together, the patterns can inspire the creation of innumerable visualizations—from simple to very complex. As the patterns are abstract, there is a great deal of flexibility as a designer proceeds through stages of design—from identifying and blending patterns, to choosing structures, to choosing representation and encoding techniques, to deciding on geometric properties. Designers can blend different patterns to devise representational structures that have different affordances for organizing information. They can blend the 14 patterns in numerous ways, using different combinations. Hence, blending the patterns provides designers with basic language-like thinking, making the design of countless representational techniques and visualizations possible. Because we are concerned mainly with complex contexts (e.g., open-ended activities, large information spaces, and heterogeneous data sources), in which static visualizations are rarely appropriate, we view visualization design in the context of human-information interaction, and we have included a discussion of interaction. Designers need to think from the outset about providing users with multiple visualizations and various complementary interactions to help them effectively work with the visualizations. Thinking about the human-visualization system as a joint cognitive system can help designers consider how to best distribute the load of information processing within the system. Using consistent terms— e.g., activity, task, interaction, event—can help with describing human-information discourse at different levels of granularity. Having an awareness of interaction patterns and techniques, along with their cognitive and epistemic utilities, can support designers in choosing and implementing interaction mechanisms in their designs. Additionally, considering interactivity—i.e., the quality of interaction—at multiple levels can help designers in operationalizing interactions in ways that will likely engender beneficial cognitive processes.

In Chapter 6, Figure 6.1 was presented to depict the different spaces in the human-visualization system. Here, we present a revised version—Figure 9.1—this time with the task and activity space also depicted. This figure is meant to summarize and bring together many of the ideas presented throughout the book. We see, for example, information space and representation space through a systems lens. The hierarchical, emergent nature of tasks and activities is depicted in the task and activity space. The overall user-visualization system is also depicted. The dashed boxes show the scope of each space. As can be seen, many of the spaces overlap—e.g., the interaction space occurs partially in the internal mental space of the user, as this is where the user's mental structures reside and where mental processes take place. Research in cognitive science has shown that while mental processes influence action—e.g., by forming intentions to act—the performance of actions also influences mental processes in significant ways. The interaction space also overlaps the representation space, as the objects of action—i.e., visual representations—exist within the representation space. The information space and computing space also overlap, as data and information are stored and processed in the computing space, and new information can be added to the information space through computational processing (e.g., statistically derived data). Finally, the task and activity space is distributed across all the other spaces. The performance of tasks and activities requires mental processing, computational processing, visual representation, and interaction with the representations. Over time, through processing in these different spaces of the joint cognitive system, tasks and activities are carried out (see Figure 6.3). Figure 9.2 depicts all of the interrelated aspects of the framework that a designer should think about when creating visualizations for an interactive human-information discourse.

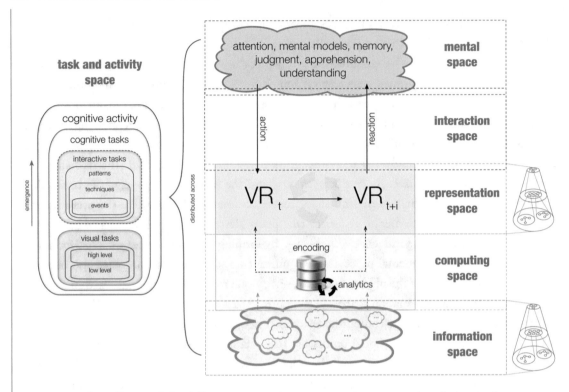

Figure 9.1: Overall view of the different spaces relevant to visual representation design in the context of a human-information discourse. VR stands for visual representation.

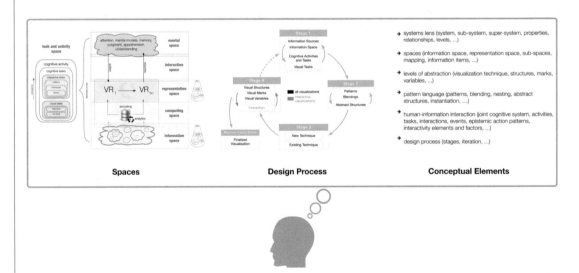

Figure 9.2: Elements of the framework that the designer can use to assist design thinking.

As mentioned previously, the core of our framework is the pattern language. We have demonstrated its utility by examining dozens of existing visualizations, and also by presenting a series of design scenarios, each one dealing with a different domain, information space, set of user tasks, and set of design requirements. We have also shown the utility of the pattern language for analyzing the structure and composition of existing visualizations. While some previous work has been focused on developing frameworks and patterns for visualization design—as described in Chapter 2—our framework differs due to the combination of the following six characteristics.

First, our framework is general—intended to be used for many types of visualizations, users, domains, tasks, and types of data and information. Unlike many previous contributions, our framework is applicable to more than typical statistical graphics and data-driven visualizations. Designers in various domains, such as STEM education, science, library science, health informatics, journalism, and others, can use this framework to support the design of novel visualizations in their domain with their intended users. Second, we are interested in supporting design thinking, specifically with respect to mapping from information space to representation space. Many previous contributions have focused on the visual representations themselves—including the rules for generating them and deconstructing them—and not on design thinking and the conceptual tools that can be used by designers to systematically analyze an information space and generate creative and novel visualizations. Third, the patterns in our framework are basic and abstract. They are fundamental building blocks, similar to letters of an alphabet, that can be combined in many different ways. Most previous work on patterns has not been conducted at this level; rather, researchers have focused on patterns for design and evaluation that capture best practices, suggest guidelines and rules, and prescribe solutions to common design problems. Fourth, because we are primarily interested in the thinking of a designer during a design process—and not in visualizations as products of the design process—we do not attempt to classify existing visualizations, as many others have done. We are concerned more with helping designers generate new visualizations than we are with classifying existing ones. We do not attempt to devise a typology of visualizations; rather, our framework is intended to support the creation of a limitless number of novel visualizations. Furthermore, we are concerned with situations in which a designer has to choose how information should be visualized and presented, and not with situations in which the designer is simply implementing something that is already specified. Fifth, we are interested in supporting visualization design in interactive contexts. In all but simple situations, being able to interact with visualizations significantly enhances the performance of a user's activity. This is especially important in the context of big data and large information spaces, which are becoming increasingly characteristic of today's activities. Finally, the sixth characteristic is that our framework is not primarily intended to help designers create simple visualizations of single datasets—e.g., a simple scatterplot. This is especially true for situations in which well-established design guidelines exist for specific data types and tasks. Previous contributions already support this type of design. Our framework is most useful in helping designers conceptualize non-simple (e.g., large, complex, dynamic, and heterogeneous) information spaces, analyze the tasks and activities of users, and generate a multiplicity of relevant and potentially complex interactive visualizations—in situations where representations are data-driven, model-driven, algorithmically generated, designer-generated, or any combination thereof—all in a holistic, principled, systematic manner.

The framework presented in this book offers a number of potential benefits for designers—whether they want to create completely novel visualizations or use and modify existing techniques. When faced with a design problem, designers need to search and think within a design space to develop a solution. Currently, the visualization design space is very large and unmanageable for most designers and situations. Our framework can act as a lens to focus a search within the design space to select existing techniques. That is, it can help reduce the search space for visualization techniques—once a designer has selected and blended a set of patterns, the number of techniques that satisfy and instantiate the particular blending is significantly lower than the number in the whole space. Future work can create design structures that function as bridges between the patterns and techniques that exist within part of the overall design space. Another benefit of the framework is that the patterns can be used to understand and create modifications of existing techniques. For example, given a set of desired patterns, and a known technique that satisfies some of the patterns, a designer can instantiate one or more patterns in structures and combine them with those of the existing technique. Additionally, the potentially overwhelming number of variations of existing techniques—such as heatmaps, treemaps, arc diagrams, and so on—can be viewed as modifications of base techniques, where one or more additional patterns are blended and instantiated to create the variant technique. Thus, especially for novice designers, the framework can act as a launching pad that helps organize and scaffold design thinking, potentially helping avoid fixation and attachment to early design ideas. Using a framework such as this one can promote reflective, critical thinking, as well as metacognitive awareness and control—all of which are known to be characteristics of successful designers.

A number of researchers have highlighted the need for a more scientific approach to visualization research and design. This is especially true at this point in time when visualizations are becoming popular in so many different domains. In other words, the design and use of visual representations is no longer of interest to only a small group of academics and designers. Moreover, computational tools in these different domains—such as intelligence analysis, bioinformatics, statistical analysis, education, and health informatics—are becoming more complex, human-centric, and highly interactive. When designing visualizations for such complex, interactive tools, designers must consider human-information discourse in a holistic manner—with user tasks, multiple visual representations, and interactions in mind. Because this is the direction in which the field is moving, we would be doing a disservice to think of a science of visualization outside of this context. At this point in time, we can no longer afford to think of visualizations as only static media—although there are situations in which static visualizations are sufficient. This is particularly the case in this age of big data, ubiquitous analytics, and complex data- and model-driven activities. It should be evident that designers need support structures (e.g., frameworks and taxonomies) that help them devise visualizations with more facility and systematicity, especially in interactive contexts. The conceptual elements in this book have been developed in an effort to address this issue. It is our hope that this contribution will spur more research and practice to move in this direction.

References

Agrawala, M., Li, W., and Berthouzoz, F. (2011). Design principles for visual communication. *Communications of the ACM*, 54(4), 60–69. DOI:10.1145/1924421.1924439. 2

Ahmed, I. and Blustein, J. (2005). Navigation in information space: how does spatial ability play a part? *Proceedings of Web-Based Communities* (IADIS), 119–125. 26

Ahlberg, C. and Shneiderman, B. (1994). Visual information seeking: Tight coupling of dynamic query filters with starfield displays. *Proceedings of the SIGCHI Conference on Human Factors in Computing Systems* (pp. 313–317). ACM. 48

Aigner, W., Miksch, S., Schumann, H., and Tominski, C. (2011). *Visualization of Time-oriented Data*. Springer Science and Business Media. DOI: 10.1007/978-0-85729-079-3. 2, 48, 124

Ainsworth, S. (1999). The functions of multiple representations. *Computers and Education*, 33(2), 131–152. DOI: 10.1016/S0360-1315(99)00029-9. 28

Ainsworth, S. (2006). DeFT: A conceptual framework for considering learning with multiple representations. *Learning and Instruction*, 16(3), 183–198. DOI: 10.1016/j.learninstruc.2006.03.001. 3

Ainsworth, S. (2008). The educational value of multiple-representations when learning complex scientific concepts. In J.K. Gilbert, M. Reiner, and M. Nakhleh (Eds.), *Visualization: Theory and Practice in Science Education*. DOI: 10.1007/978-1-4020-5267-5_9. 4

Alexander, C. (1964). *Notes on the Synthesis of Form*. Harvard University Press. Cambridge, MA. 15, 19

Alexander, C., Ishikawa, S., Silverstein, M., Jacobson, M., Fiksdahl-King, I., and Angel, S. (1977). *A Pattern Language: Towns, Buildings, Construction*. NY: Oxford University Press. 19, 50

Alexander, C. (1979). *The Timeless Way of Building*. Oxford University Press. 47

Allen, B. L. (1998). Visualization and cognitive abilities. Visualizing subject access for 21st century information resources [papers presented at the 1997 Clinic on Library Applications of Data Processing, March 2–4, 1997 Urbana-Champaign]. 26

Alper, B., Riche, N. H., Ramos, G., and Czerwinski, M. (2011). Design study of linesets, a novel set visualization technique. *IEEE Transactions on Visualization and Computer Graphics*, 17(12), 2259–2267. DOI: 10.1109/TVCG.2011.186. 36

Alsallakh, B., Aigner, W., Miksch, S., and Hauser, H. (2013). Radial sets: Interactive visual analysis of large overlapping sets. *IEEE Transactions on Visualization and Computer Graphics*, 19(12), 2496–2505. DOI: 10.1109/TVCG.2013.184. 103

Amar, R., Eagan, J., and Stasko, J. (2005). Low-level components of analytic activity in information visualization. *IEEE Symposium on Information Visualization (INFOVIS 2005)*, (pp. 111–117). IEEE. DOI: 10.1109/INFOVIS.2005.24. 110

Arcavi, A. (2003). The role of visual representations in the learning of mathematics. *Educational Studies in Mathematics*, 52(3), 215–241. DOI: 10.1023/A:1024312321077. 28

Baldonado, W. M., Woodruff, A., and Kuchinsky, A. (2000). Guidelines for using multiple views in information visualization. *Proceedings of the Working Conference on Advanced Visual Interfaces (AVI '00)*. DOI: 10.1145/345513.345271. 4

Ball, L. J., Evans, J. S. B., and Dennis, I. (1994). Cognitive processes in engineering design: a longitudinal study. *Ergonomics*, 37(11), 1753–1786. DOI: 10.1080/00140139408964950. 21

Bayle, E., Bellamy, R., Casaday, G., Erickson, T., Fincher, S., Grinter, B., Gross, B., Lehder, D., Marmolin, H., Moore, B., Potts, C., Skousen, G., and Thomas, J. (1998). Putting it all together: Toward a pattern language for interaction. *SIGCHI Bulletin*, 30(1), 17–33. DOI: 10.1145/280571.280580. 19, 20, 47

Benton, M. J. (1998). The quality of the fossil record of vertebrates. In S.K. Donovan and C.R.C. Paul (Eds.), *The Adequacy of the Fossil Record*. 269–303. Wiley, New York. 89

Benyon, D. (2005). Information space. In C. Ghaoui (Ed.), *Encyclopedia of Human Computer Interaction*, 344–347. Idea Group Reference. 26

Bertin, J. (1967). Sémiologie Graphique. Les diagrammes, les réseaux, les cartes. With Marc Barbut [et al.]. Paris: Gauthier-Villars. (Translation 1983. Semiology of Graphics by William J. Berg.) 9, 11, 37, 127

Bodner, G. M. and Domin, D. S. (2000). Mental models: The role of representations in problem solving in chemistry. *University Chemistry Education*, 4(1), 24–30. 28

Bohm, D. and Peat, D. (1987). *Science, Order, and Creativity*. Bantam Books. 16

Boisot, M. and Canals, A. (2007). Data, information, and knowledge: Have we got it right?. In M. Boisot, I. MacMillan, and K. Han (Eds.), *Explorations in Information Space: Knowledge, Agents, and Organization*. NY:Oxford University Press. DOI: 10.1093/acprof: oso/9780199250875.003.0002. 26, 121

Borchers, J. O. (2001). A pattern approach to interaction design. *Ai and Society*, 15(4), 359–376. DOI: 10.1007/BF01206115. 19

Börner, K. (2010). Atlas of science: Visualizing what we know. MIT Press. 124

Börner, K. (2015). Atlas of knowledge: Anyone can map. MIT Press. 37, 124, 127

Boroditsky, L. (2001). Does language shape thought? English and Mandarin speakers' conceptions of time. *Cognitive Psychology* 43(1), 1–22. DOI: 10.1006/cogp.2001.0748. 18

Bostock, M. and Heer, J. (2009). Protovis: A graphical toolkit for visualization. *IEEE Transactions on Visualization and Computer Graphics*, 15(6), 1121–1128. DOI: 10.1109/TVCG.2009.174. 15

Bostock, M., Ogievetsky, V. and Heer, J. (2011). D^3 data-driven documents. *Visualization and Computer Graphics, IEEE Transactions on*, 17(12), 2301–2309. DOI: 10.1109/TVCG.2011.185. 15

Brehmer, M. and Munzner, T. (2013). A multi-level typology of abstract visualization tasks. *Visualization and Computer Graphics, IEEE Transactions on*, 19(12), 2376–2385. DOI: 10.1109/TVCG.2013.124. 110, 121

Brey, P. (2005). The epistemology and ontology of human-computer interaction. *Minds and Machines*, 15(3–4), 383–398. DOI:10.1007/s11023-005-9003-1. 4

Burch, M. and Diehl, S. (2008). TimeRadarTrees: Visualizing dynamic compound digraphs. *Computer Graphics Forum*, 27(3), 823–830. DOI: 10.1111/j.1467-8659.2008.01213.x. 105

Burkhard, R. A. (2004). Learning from architects: the difference between knowledge visualization and information visualization. In *Information Visualisation, 2004. IV 2004. Proceedings. Eighth International Conference on* (pp. 519–524). IEEE. DOI: 10.1109/IV.2004.1320194. 28

Burkhard, R. A. (2005). Toward a framework and a model for knowledge visualization: Synergies between information and knowledge visualization. In *Knowledge and Information Visualization* (pp. 238–255). Springer Berlin Heidelberg. DOI: 10.1007/11510154_13. 3

Cabré, M.T. (1999). *Terminology: Theory, Methods and Applications*. Philadelphia: John Benjamins DOI: 10.1075/tlrp.1. 17

Cabré, M.T. (2003). Theories of terminology: Their description, prescription and explanation. *Terminology*, 9(2), 163–199. DOI: 10.1075/term.9.2.03cab. 18

Cairo, A. (2012). *The Functional Art: An Introduction to Information Graphics and Visualization*. New Riders. 48, 127

Card, S. K., Mackinlay, J. D., and Shneiderman, B. (1999). *Readings in Information Visualization: Using Vision to Think*. Morgan Kaufmann Publishers. 3, 37, 121, 124, 127

Carpendale, M. S. T. (2003). Considering visual variables as a basis for information visualisation. Technical Report 2001-693-16, Department of Computer Science, University of Calgary, Calgary, Canada. 9, 127

Carroll, J. M. (2000). *Making Use: Scenario-based Design of Human-computer Interactions*. MIT Press. DOI: 10.1145/347642.347652. 119

Carswell, C. M. (1992). Choosing specifiers: An evaluation of the basic tasks model of graphical perception. *Human Factors: The Journal of the Human Factors and Ergonomics Society*, 34(5), 535–554. 45, 111

Casakin, H.P. (2006). Assessing the use of metaphors in the design process. *Environment and Planning B: Planning and Design*, 33(2), 253–268. DOI: 10.1068/b3196. 26

Casakin, H.P. (2007). Metaphors in design problem solving: Implications for creativity. *International Journal of Design*, 1(2), 21–33. 26

Chen, C. (2010). Information Visualization. *Wiley Interdisciplinary Reviews: Computational Statistics* 2.4, 387–403. DOI: 10.1002/wics.89. 26

Chen, H. (2004). Toward design patterns for dynamic analytical data visualization. *Proceedings of SPIE-IS&T Electronic Imaging* (Vol. 5295, pp. 75–86). SPIE and IS&T. 19

Chen, M., Ebert, D., Hagen, H., Laramee, R.S., van Liere, R., Ma, K.-L., Ribarsky, W., Scheuermann, G., and Silver, D. (2009). Data, Information, and Knowledge in Visualization. *Computer Graphics and Applications*, 29(1), 12–19. DOI: 10.1109/MCG.2009.6. 26

Cheng, P. C. H. (2002). Electrifying diagrams for learning: principles for complex representational systems. *Cognitive Science*, 26(6), 685–736. DOI: 10.1207/s15516709cog2606_1. 3

Chi, E. H. (2000). A taxonomy of visualization techniques using the data state reference model. *Information Visualization, 2000. InfoVis 2000. IEEE Symposium on* (pp. 69–75). IEEE. DOI: 10.1109/INFVIS.2000.885092. 2, 3, 124

Clark, A. (1998). Time and mind. *Journal of Philosophy*, 95(7), 354–376. DOI:10.2307/2564539. 4

Cleveland, W.S. (1985). *The Elements of Graphing Data*. Wadsworth, Inc. 10, 11

Cleveland, W.S. and McGill, R. (1984). Graphical perception: Theory, experimentation, and application to the development of graphical methods. *Journal of the American Statistical Association*, 79(387), 531–554. DOI: 10.1080/01621459.1984.10478080. 3, 10, 11, 43, 44, 127

Cleveland, W.S. and McGill, R. (1986). An experiment in graphical perception. *International Journal of Man-Machine Studies*, 25(5), 491–500. DOI: 10.1016/S0020-7373(86)80019-0. 10, 127

Cooper, A., Reimann, R., Cronin, D. and Noessel, C. (2014). *About Face: The Essentials of Interaction Design*, 4th Ed. John Wiley & Sons. 109

Coopmans, C., Vertesi, J., Lynch, M., and Woolgar, S. (2014). *Representation in Scientific Practice Revisited*. MIT Press. DOI: 10.7551/mitpress/9780262525381.001.0001. 124

Corbett, A. T., and Anderson, J. R. (2001). Locus of feedback control in computer-based tutoring: Impact on learning rate, achievement and attitudes. *Proceedings of the SIGCHI Conference on Human Factors in Computing Systems* (pp. 245–252). ACM. DOI: 10.1145/365024.365111. 116

Cross, N. (1981). Design method and scientific method. *Design Studies*, 2(4), 195–201. DOI: 10.1016/0142-694X(81)90050-8. 15

Cross, N. (2004). Expertise in design: an overview. Design Studies, 25, 427–441. DOI:10.1016/j.destud.2004.06.002. 17, 21

Cross, N. (2011). *Design Thinking: Understanding how Designers Think and Work*. Berg, Oxford, UK. 15, 16

Cuoco, A. A. and Curcio, F. R. (Eds., 2001). *The Roles of Representation in School Mathematics*. Reston, Virginia: National Council of Teachers of mathematics. 124

Dade-Robertson, M. (2011). *The Architecture of Information: Architecture, Interaction Design and the Patterning of Digital Information*. Routledge. 26

Dearden, A.M. and Finlay, J. (2006). Pattern languages in HCI: A critical review. *Human Computer Interaction*, 21(1), 49–102. DOI: 10.1207/s15327051hci2101_3. 20, 47

Devesa S. S., Grauman D. G., Blot W. J., Pennello G., Hoover R. N., Fraumeni J. F. Jr. (1999). *Atlas of Cancer Mortality in the United States, 1950-94*. Washington, D.C.: US Govt Print Off; [NIH Publ No. (NIH) 99–4564]. 75

Diaper, D. and Stanton, N. (Eds., 2003). *The Handbook of Task Analysis for Human-computer Interaction*. CRC Press. 121

Dillon, A. (2000). Spatial-semantics: How users derive shape from information space. *Journal of the American Society for Information Science*, 51(6), 521–528. DOI: 10.1002/(SICI)1097-4571(2000)51:6<521::AID-ASI4>3.0.CO;2-5. 26

Dix, A. and Ellis, G. (1998). Starting simple—Adding value to static visualisation through simple interaction. *Proceedings of the Working Conference on Advanced Visual Interfaces (AVI)*. DOI:10.1145/948496.948514. 4

Dörk, M., Carpendale, S., and Williamson, C. (2012). Visualizing explicit and implicit relations of complex information spaces. *Information Visualization*, 11(1), 5–21. DOI: 10.1177/1473871611425872. 27, 48

Eilam, B. and Gilbert, J. K. (Eds., 2014). *Science Teachers' Use of Visual Representations*. Springer. DOI: 10.1007/978-3-319-06526-7. 124

Einstein, A. (1954). *Ideas and Opinions*. Crown Publishers. 16

Elmqvist, N., Moere, A. V., Jetter, H. C., Cernea, D., Reiterer, H., and Jankun-Kelly, T. J. (2011). Fluid interaction for information visualization. *Information Visualization*, 10(4), 327–340. DOI: 10.1177/1473871611413180. 4, 19, 118, 128

Elmqvist, N. and Yi, J. S. (2015). Patterns for visualization evaluation. *Information Visualization*, 14(3), 250–269. DOI: 10.1177/1473871613513228. 20

Engelhardt, Y. (2002). The language of graphics. Doctoral Dissertation. Institute for Logic, Language and Computation, University of Amsterdam. 2, 11, 12, 14

Engelhardt, Y. (2006). Objects and Spaces: The visual language of graphics. In D. Barker-Plummer et al. (Eds.), Diagrams 2006, LNAI 4045, pp. 104–108. DOI: 10.1007/11783183_13. 11, 12

Few, S. (2004). *Show Me the Numbers: Designing Tables and Graphs to Enlighten*. Analytics Press. 2

Few, S. (2009). *Now You See It: Simple Visualization Techniques for Quantitative Analysis*. Analytics Press. 2, 127

Few, S. (2013). *Information Dashboard Design: Displaying Data for At-a-glance Monitoring* (2nd Ed.). Analytics Press. 124, 127

Fincher, S. (1999). Analysis of Design: An Exploration of Patterns and Pattern Languages for Pedagogy. *Journal of Computers in Mathematics and Science Teaching: Special Issue on Computer Science Education*, 18 (3), pp. 331–348. 47, 48

Fisher, B., Green, T. M., and Arias-Hernández, R. (2011). Visual analytics as a translational cognitive science. *Topics in Cognitive Science*, 3(3), 609–625. DOI: 10.1111/j.1756-8765.2011.01148.x. 107

Fricke, G. (1996). Successful individual approaches in engineering design. *Research in Engineering Design*, 8(3), 151–165. DOI: 10.1007/BF01608350. 21

Gabriel R. P. (1996). *Patterns of Software: Tales from the Software Community*. Oxford University Press. 48

Gamma, E., Helm, R., Johnson, R., and Vlissides, J. (1995). *Design Patterns: Elements of Reusable Object-oriented Software*. Addison-Wesley. 19

Gapminder. www.gapminder.org. 48

Gedenryd, H. (1998). How Designers Think. Doctoral Dissertation. Lund University. 15, 16

Gerjets, P., Imhof, B., Kühl, T., Pfeiffer, V., Scheiter, K., and Gemballa, S. (2010). Using Static and Dynamic Visualizations to Support the Comprehension of Complex Dynamic Phenomena in the Natural Sciences. Use of External Representations in Reasoning and Problem Solving: Analysis and Improvement. London:Routledge, 153–168. 28

Gero, J. S. (2000). Computational models of innovative and creative design processes. *Technological Forecasting and Social Change*, 64(2–3), 183–196. DOI: 10.1016/S0040-1625(99)00105-5. 26

Glasgow, J., Narayanan, N. H. and Chandrasekaran, B., Eds. (1995) *Diagrammatic Reasoning: Cognitive and Computational Perspectives*. MIT Press. 124

Gotz, D., Lu, J., Kissa, P., Cao, N., Qian, W. H., Liu, S. X., and Zhou, M. X. (2010). HARVEST: an intelligent visual analytic tool for the masses. *Proceedings of the First International Workshop on Intelligent Visual Interfaces for Text Analysis* (pp. 1–4). ACM. DOI: 10.1145/2002353.2002355. 48

Green, T. M. and Fisher, B. (2011). The personal equation of complex individual cognition during visual interface interaction. In A. Ebert, A. Dix, N.D. Gershon, and M. Pohl (Eds.), *Human Aspects of Visualization* (pp. 38–57). Springer Berlin Heidelberg. DOI: 10.1007/978-3-642-19641-6_4. 118, 128

Guindon, R. (1990). Designing the design process: Exploiting opportunistic thoughts. *Human-Computer Interaction*, 5(2), 305–344. DOI: 10.1207/s15327051hci0502&3_6. 21

Hackos, J. T. and Redish, J. (1998). *User and Task Analysis for Interface Design*. John Wiley and Sons Inc. 121

Hansen, C. D. and Johnson, C. R. (2005). *The Visualization Handbook*. Academic Press. 36, 121, 124

Harris, R. L. (1999). *Information Graphics: A Comprehensive Illustrated Reference*. Oxford University Press. 2, 11, 12, 14, 36, 48, 124

Hatcher, A. (2011). *Topology of Numbers*. Unpublished manuscript, in preparation. 93

Havre, S., Hetzler, B., and Nowell, L. (2000). ThemeRiver: Visualizing theme changes over time. *IEEE Symposium on Information Visualization (InfoVis 2000)*. pp. 115–123. DOI: 10.1109/infvis.2000.885098. 98

Hearst, M. A. (1995). TileBars: visualization of term distribution information in full text information access. *Proceedings of the SIGCHI Conference on Human Factors in Computing Systems* (pp. 59–66). ACM Press/Addison-Wesley Publishing Co. DOI: 10.1145/223904.223912. 35

Heer, J. and Agrawala, M. (2006). Software design patterns for information visualization. *IEEE Transactions on Visualization and Computer Graphics*, 12(5), 853–860. DOI: 10.1109/TVCG.2006.178. 19

Heer, J. and Bostock, M. (2010). Crowdsourcing graphical perception: using mechanical turk to assess visualization design. *Proceedings of the SIGCHI Conference on Human Factors in Computing Systems* (pp. 203–212). ACM. DOI: 10.1145/1753326.1753357. 9

Heer, J., Bostock, M., and Ogievetsky, V. (2010). A tour through the visualization zoo. *Commun. ACM*. Retrieved from http://queue.acm.org/detail.cfm?id=1743567. DOI: 10.1145/1743546.1743567. 2, 48, 124

Heer, J. and Boyd, D. (2005). Vizster: Visualizing online social networks. *IEEE Symposium on Information Visualization (INFOVIS 2000)*, (pp. 32–39). DOI: 10.1109/INFOVIS.2005.39. 48

Heer, J. and Shneiderman, B. (2012). Interactive dynamics for visual analysis. *ACM Queue*, 10(2), 30 pp. DOI: 10.1145/2133416.2146416. 27, 118, 128

Hegarty, M. (2011). The cognitive science of visual-spatial displays: Implications for design. *Topics in Cognitive Science*, 3(3), 446–474. DOI: 10.1111/j.1756-8765.2011.01150.x. 3

Hegarty, M. and Kozhevnikov, M. (1999). Types of visual–spatial representations and mathematical problem solving. *Journal Educational Psychology*, 91(4), 684. DOI: 10.1037/0022-0663.91.4.684. 28

Henry, N., Fekete, J. D., and McGuffin, M. J. (2007). NodeTrix: a hybrid visualization of social networks. *Visualization and Computer Graphics, IEEE Transactions on*, 13(6), 1302–1309. DOI: 10.1109/TVCG.2007.70582. 48, 92

Hentschel, K. (2002). *Mapping the Spectrum: Techniques of Visual Representation in Research and Teaching*. Oxford University Press. DOI: 10.1093/acprof:oso/9780198509530.001.0001. 124

Hollan, J., Hutchins, E., and Kirsh, D. (2000). Distributed cognition: toward a new foundation for human-computer interaction research. *ACM Transactions on Computer-Human Interaction (TOCHI)*, 7(2), 174–196. DOI: 10.1145/353485.353487. 108

Hutchins, E. (1995). *Cognition in the Wild*. MIT press. 2, 108

Hutchins, E. L., Hollan, J. D., and Norman, D. A. (1985). Direct manipulation interfaces. *Human–Computer Interaction*, 1(4), 311–338. DOI: 10.1207/s15327051hci0104_2. 116

Iacob, C. (2011). A design pattern mining method for interaction design. *Proceedings of the 3rd ACM SIGCHI Symposium on Engineering Interactive Computing Systems (EICS '11)*, 217–222. DOI: 10.1145/1996461.1996523. 48

Iliinsky, N. P. N. (2003). Generation of complex diagrams: How to make lasagna instead of spaghetti. Unpublished Masters Thesis. Department of Technical Communication, University of Washington. 96

Javed, W. and Elmqvist, N. (2012). Exploring the design space of composite visualization. *IEEE Pacific Visualization Symposium (PacificVis)* (pp. 1–8). DOI: 10.1109/pacificvis.2012.6183556. 3, 19

Johnson, J. (2013). *Designing with the Mind in Mind: Simple Guide to Understanding User Interface Design Guidelines*. Elsevier. 119

Jonassen, D. H. (1995). Computers as cognitive tools: Learning with technology, not from technology. *Journal of Computing in Higher Education*, 6(2), 40–73. DOI: 10.1007/BF02941038. 108

Jonassen, D. H., Beissner, K., and Yacci, M. (Eds., 1993). *Structural Knowledge: Techniques for Representing, Conveying, and Acquiring Structural Knowledge*. Hillsdale, NJ: Lawrence Erlbaum Associates. 36, 124

Jonassen, D. H., Tessmer, M., and Hannum, W. H. (1998). *Task Analysis Methods for Instructional Design*. Routledge. 122

Jong, D. T., Ainsworth, S., Dobson, M., Hulst, A. van der, Levonen, J., Reimann, P., Sime, J.A., Someren, M. van, Spada, H., and Swaak, J. (1998) Acquiring knowledge in science and mathematics: the use of multiple representations in technology based learning environments. In: M. W. van Someren (Ed.), *Learning with Multiple Representations. Advances in Learning and Instruction Series*. Oxford:Elsevier Science, (pp. 9–41). 28

Juchem, C., Muller-Bierl, B., Schick, F., Logothetis, N. K., and Pfeuffer, J. (2006). Combined passive and active shimming for in vivo MR spectroscopy at high magnetic fields. *Journal of Magnetic Resonance*, 183(2), 278–289. DOI: 10.1016/j.jmr.2006.09.002. 97

Kavakli, M. and Gero, J. S. (2002). The structure of concurrent cognitive actions: a case study on novice and expert designers. *Design Studies*, 23(1), 25–40. DOI: 10.1016/S0142-694X(01)00021-7. 21

Keim, D. A., Kohlhammer, J., Ellis, G., and Mansmann, F. (Eds., 2010). *Mastering the Information Age-solving Problems with Visual Analytics*. Eurographics Association. 27

Kieras, D., Meyer, D., and Ballas, J. (2001). Toward demystification of direct manipulation: Cognitive modeling charts the gulf of execution. *Proceedings of the SIGCHI Conference on Human Factors in Computing Systems* (pp. 128–135). ACM. DOI: 10.1145/365024.365069. 116

Kirsh, D. (1997). Interactivity and multimedia interfaces. *Instructional Science*, 25(2), 79–96. DOI: 10.1023/A:1002915430871. 4, 108, 118

Kirsh, D. (2005). Metacognition, distributed cognition and visual design. *Cognition, Education, and Communication Technology*, 147–180. 4, 108

Kirsh, D. (2010). Thinking with external representations. *Ai and Society*, 25(4), 441–454. DOI: 10.1007/s00146-010-0272-8. 2, 108

Kirsh, D. (2013). Embodied cognition and the magical future of interaction design. *ACM Transactions on Computer-Human Interaction (TOCHI)*, 20(1), 3. DOI: 10.1145/2442106.2442109. 4, 118

Kirsh, D. (2014). The importance of chance and interactivity in creativity. *Pragmatics & Cognition*, 22(1), 5–26. DOI: 10.1075/pc.22.1.01kir. 17

Kirsh, D. and Maglio, P. (1994). On distinguishing epistemic from pragmatic action. *Cognitive Science*, 18(4), 513–549. DOI: 10.1207/s15516709cog1804_1. 4, 114

Knauff, M. and Wolf, A. G. (2010). Complex cognition: the science of human reasoning, problem-solving, and decision-making. *Cognitive Processing*, 11(2), 99–102. DOI: 10.1007/s10339-010-0362-z. 110

Koenig, A. (1995). Patterns and Antipatterns. *Journal of Object-Oriented Programming*, 8(1): 46–48. 50

Kosslyn, S. M. (2006). *Graph Design for the Eye and Mind*. Oxford University Press. DOI: 10.1093/acprof:oso/9780195311846.001.0001. 3

Kozma, R. (2003). The material features of multiple representations and their cognitive and social affordances for science understanding. *Learning and Instruction*, 13(2), 205–226. DOI:10.1016/S0959-4752(02)00021-X. 4

Kreuseler, M. and Schumann, H. (2002). A flexible approach for visual data mining. *IEEE Transactions on Visualization and Computer Graphics*, 8(1), 39–51. DOI: 10.1109/2945.981850. 121

Krippendorff, K. (2006). *The Semantic Turn: A New Foundation for Design*. Boca Raton: Taylor and Francis. DOI: 10.4324/9780203299951. 20

Krum, R. (2013). *Cool Infographics: Effective Communication with Data Visualization and Design*. John Wiley and Sons. 124

Kutz, D. O. (2004). Examining the evolution and distribution of patent classifications. *Proceedings of the Eighth International Conference on Information Visualisation (IV 2004)*, (pp. 983–988). IEEE. DOI: 10.1109/iv.2004.1320261. 93

van Labeke, N. V. and Ainsworth, S. E. (2001). Applying the DeFT framework to the design of multi-representational instructional simulations. *AIED'01 - 10th International Conference on Artificial Intelligence in Education*, 19-23 May 2001, San Antonio, Texas, IOS Press, pp. 314–321. 48

Lajoie, S. P. and Derry, S. J., Eds. (1993). *Computers as Cognitive Tools*. Lawrence Erlbaum Associates, Inc. 108

Lakoff, G. and Johnson, M. (1996). *Metaphors We Live By*. University of Chicago Press. 26

Lankow, J., Ritchie, J., and Crooks, R. (2012). *Infographics: The Power of Visual Storytelling*. John Wiley and Sons. 124

Lanzilotti, R., Ardito, C., Costabile, M. F., and De Angeli, A. (2011). Do patterns help novice evaluators? *A comparative study. International Journal of Human-computer Studies*, 69(1), 52–69. DOI: 10.1016/j.ijhcs.2010.07.005. 20

Larkin, J. H. and Simon, H. A. (1987). Why a diagram is (sometimes) worth ten thousand words. *Cognitive Science*, 11(1), 65–100. DOI: 10.1111/j.1551-6708.1987.tb00863.x. 4, 9

Laszlo, E. (1972). *The Systems View of the World: The Natural Philosophy of the New Developments in the Sciences*. NY: George Braziller. 23

Levinson, S. C. (2003). *How Deep Are Effects of Language on Thought?* Cambridge, UK: Cambridge University Press. 18

Li, S., Crouser, R. J., Griffin, G., Gramazio, C., Schulz, H. J., Childs, H., and Chang, R. (2015). Exploring hierarchical visualization designs using phylogenetic trees. *IS&T/SPIE Electronic Imaging* (pp. 939709–939709). International Society for Optics and Photonics. 2

Liang, H, Parsons, P, Wu, H, and Sedig, K (2010). An exploratory study of interactivity in visualization tools: 'Flow' of interaction. *Journal of Interactive Learning Research*, 19(1), 103–120. 116, 118

Liang, H. N., and Sedig, K. (2010). Role of interaction in enhancing the epistemic utility of 3D mathematical visualizations. *International Journal of Computers for Mathematical Learning*, 15(3), 191–224. DOI: 10.1007/s10758-010-9165-7. 4, 118

Lidwell, W., Holden, K., and Butler, J. (2010). *Universal Principles of Design, Revised and Updated: 125 Ways to Enhance Usability, Influence Perception, Increase Appeal, Make Better Design Decisions, and Teach through Design*. Rockport Publishers, Inc. Beverly, MA. 119

Lima, M. (2011). *Visual Complexity: Mapping Patterns of Information*. Princeton Architectural Press. 2, 36, 48, 124

Lima, M. (2014). *The Book of Trees: Visualizing Branches of Knowledge*. Princeton Architectural Press. 2, 36, 124

Livnat, Y., Agutter, J., Moon, S., and Foresti, S. (2005). Visual correlation for situational awareness. *IEEE Symposium on Information Visualization (INFOVIS 2005)*, pp. 95–102. IEEE. DOI: 10.1109/infvis.2005.1532134. 101

Liu, Z. and Heer, J. (2014). The effects of interactive latency on exploratory visual analysis. *IEEE Transactions on Visualization and Computer Graphics*, 20(12), 2122–2131. DOI: 10.1109/TVCG.2014.2346452. 116

Liu, Z., Nersessian, N. J., and Stasko, J. T. (2008). Distributed cognition as a theoretical framework for information visualization. *IEEE Transactions on Visualization and Computer Graphics*, 14(6), 1173–1180. DOI: 10.1109/TVCG.2008.121. 107, 108

Lohse, G. L., Biolsi, K., Walker, N., and Rueter, H. H. (1994). A classification of visual representations. *Communications of the ACM*, 37(12), 36–49. DOI:10.1145/198366.198376. 2, 9, 12, 14, 48

MacEachren, A.M. (1995). *How Maps Work: Representation, Visualization, and Design*. New York: Guilford Press. 3, 9, 10, 12, 36, 121, 124

Mackinlay, J. (1986). Automating the design of graphical presentations of relational information. *ACM Transactions on Graphics*, 5(2), 110–141. DOI: 10.1145/22949.22950. 2, 9, 10, 12, 44, 45, 127

Maglio, P.P. and Matlock, T. (1999). The conceptual structure of information space. In A.J. Munro, K. Höök, and D. Benyon (Eds.), *Social Navigation of Information Space*. Springer. DOI: 10.1007/978-1-4471-0837-5_9. 26

Malcolm, G. (2004). *Multidisciplinary Approaches to Visual Representations and Interpretations* (Vol. 2). Elsevier. 36, 124

Markman, A. B. (1999). *Knowledge Representation*. Mahwah, NJ: Lawrence Erlbaum Associates. 28, 36, 124

Mazza, R. (2009). *Introduction to Information Visualization*. Springer Science and Business Media. 48, 127

Meadows, D. H. (2008). *Thinking in Systems: A Primer*. Chelsea Green Publishing. 121

Meirelles, I. (2013). *Design for Information: An Introduction to the Histories, Theories, and Best Practices Behind Effective Information Visualizations*. Rockport. 2, 36, 48, 124

Meszaros, G. (1996). Patterns for Decision Making in Architectural Design. *Addendum to the proceedings of OOPSLA '95 OOPS Messenger* 6(4) 132–137. 48

Moktefi, A. and Shin, S.-J. (Eds., 2013). *Visual Reasoning with Diagrams*. Springer. 36, 124

Moody, J. and Mucha, P. J. (2013). Portrait of political party polarization. *Network Science*, 1(01), 119–121. DOI: 10.1017/nws.2012.3. 99

Moreno, R., Ozogul, G., and Reisslein, M. (2011). Teaching with concrete and abstract visual representations: Effects on students' problem solving, problem representations, and learning perceptions. *Journal of Educational Psychology*, 103(1), 32. DOI: 10.1037/a0021995. 28

Morey, J. and Sedig, K. (2004). Adjusting degree of visual complexity: An interactive approach for exploring four-dimensional polytopes. *The Visual Computer: International Journal of Computer Graphics*, 20, 1–21. DOI: 10.1007/s00371-004-0259-x. 48, 118

Morey, J., Sedig, K., and Mercer, R. E. (2001). Interactive metamorphic visuals: Exploring polyhedral relationships. *Proceedings Fifth International Conference on Information Visualisation* (p. 0483). IEEE. DOI: 10.1109/IV.2001.942100. 4, 118

Munzner, T. (2009). A nested model for visualization design and validation. *IEEE Transactions on Visualization and Computer Graphics*, 15(6), 921–928. DOI: 10.1109/TVCG.2009.111. 3, 119

Munzner, T. (2015). *Visualization Analysis and Design*. CRC Press. 2, 37, 121, 127

Narayanan, N. H. and Hegarty, M. (2002). Multimedia design for communication of dynamic information. *International Journal Human-Computer Studies*, 57(4), 279–315. DOI: 10.1006/ijhc.2002.1019. 3

Nardi, B. A. (1996). Studying context: A comparison of activity theory, situated action models, and distributed cognition. In B.A. Nardi (Ed.), *Context and Consciousness: Activity Theory and Human-computer Interaction*, (pp. 69–102). MIT Press. 108

Nelson, H. G. and Stolterman, E. (2012). *The Design Way: Intentional Change in an Unpredictable World* (2nd ed.). MIT Press. 16, 119

Norman, D. A. (1999). Affordance, conventions, and design. *Interactions*, 6(3), 38–43. DOI: 10.1145/301153.301168. 119

Norman, D. A. (2005). Human-centered design considered harmful. *Interactions*, 12(4), 14–19. DOI: 10.1145/1070960.1070976. 109

Norman, D. A. (2013). *The Design of Everyday Things*. Revised and expanded edition. Basic books.

Norman, D. A., and Draper, S. W. (1986). *User Centered System Design: New Perspectives on Human-Computer Interaction*. Lawrence Erlbaum Associates, Inc. Hillsdale, NJ. 119

Nowell, L.T. (1997). Graphical encoding for information visualization: Using icon color, shape, and size to convey nominal and quantitative data. Doctoral Dissertation. Virginia Polytechnic Institute and State University. DOI: 10.1145/1120212.1120256. 9, 127

Parsons, P. and Sedig, K. (2013a). Common visualizations: Their cognitive utility. In *Handbook of Human Centric Visualization*. New York:Springer (pp. 671–691). 2

Parsons, P. and Sedig, K. (2013b). Adjustable properties of visual representations: Improving the quality of human-information interaction. *Journal of the Association for Information Science and Technology*, 65(3), 455–482. DOI:10.1002/asi.23002. 3, 4, 13, 118, 128

Parsons, P. and Sedig, K. (2013c). Distribution of information processing while performing complex cognitive activities with visualization tools. In *Handbook of Human Centric Visualization* (pp. 693–715). New York:Springer. 107, 108, 118

Pauwels, L. (2006). *Visual Cultures of Science: Rethinking Representational Practices in Knowledge Building and Science Communication*. UPNE. 36, 124

Peterson, D. (Ed., 1996). *Forms of Representation*. Exeter, UK:Intellect Books. 28, 36, 124

Pike, W., Stasko, J., Chang, R., and O'Connell, T. (2009). The science of interaction. *Information Visualization*, 8(4), 263-274. DOI:10.1057/ivs.2009.22. 4, 118, 128

Pirolli, P. and Card, S. (2005). The sensemaking process and leverage points for analyst technology as identified through cognitive task analysis. *Proceedings of International Conference on Intelligence Analysis*, 6pp. 26, 122

Potter, S. S., Roth, E. M., Woods, D. D., and Elm, W. C. (2000). Bootstrapping multiple converging cognitive task analysis techniques for system design. In J.M. Schraagen, S.F. Chipman, and V.L. Shalin (Eds.), *Cognitive Task Analysis* (pp. 317–340). Lawrence Erlbaum Associates, Inc. 122

Rao, R. and Card, S. K. (1994). The table lens: merging graphical and symbolic representations in an interactive focus+ context visualization for tabular information. *Proceedings of the SIGCHI Conference on Human Factors in Computing Systems*. ACM. (pp. 318–322). DOI: 10.1145/191666.191776. 48

Reeves, T. C., Benson, L., Elliott, D., Grant, M., Holschuh, D., Kim, B., ... and Loh, S. (2002). Usability and Instructional Design Heuristics for E-Learning Evaluation. *Proceedings of ED-MEDIA 2002 World Conference on Educational Multimedia, Hypermedia and Telecommunications*, (8 pp.). 119

Rensink, R. A. (2013). On the prospects for a science of visualization. In W. Huang (Ed.), *Handbook of Human Centric Visualization* (pp. 147–175). Springer. 3

Rind, A., Aigner, W., Wagner, M., Miksch, S., and Lammarsch, T. (2015). Task Cube: A three-dimensional conceptual space of user tasks in visualization design and evaluation. *Information Visualization*. DOI: 10.1177/1473871615621602. 109, 110, 122

Roberts, J. C. (2007). State of the art: Coordinated and multiple views in exploratory visualization. *Proceedings of the 5th International Conference on Coordinated and Multiple Views in Exploratory Visualization (CMV '07)*. Retrieved from http://ieeexplore.ieee.org/xpls/abs_all.jsp?arnumber=4269947. DOI: 10.1109/CMV.2007.20. 4

Roth, R. E. (2013). An empirically-derived taxonomy of interaction primitives for interactive cartography and geovisualization. *IEEE Transactions on Visualization and Computer Graphics*, 19(12), 2356–2365. DOI: 10.1109/TVCG.2013.130. 118, 128

Roth, S. F., and Mattis, J. (1990). Data characterization for intelligent graphics presentation. *Proceedings of the SIGCHI Conference on Human Factors in Computing Systems* (pp. 193–200). ACM. DOI: 10.1145/97243.97273. 121

Rowe, P. G. (1987). *Design Thinking*. MIT Press. 21

Sagan. C. (1990). Why We Need to Understand Science. *The Skeptical Inquirer*, 14(3). 16

Salingaros, N.A. (1999). Architecture, patterns, and mathematics. *Nexus Network Journal*, 1, 75–85. DOI: 10.1007/s00004-998-0006-0. 47

Salingaros, N.A. (2000). The structure of pattern languages. *Architectural Research Quarterly*, 4, 149–161. DOI: 10.1017/S1359135500002591. 17, 47

Salomon, G. (Ed., 1993). Distributed Cognitions: Psychological and Educational Considerations. Cambridge University Press. 108

Salomon, G., Perkins, D. N., and Globerson, T. (1991). Partners in cognition: Extending human intelligence with intelligent technologies. *Educational Researcher*, 20(3), 2–9. DOI: 10.3102/0013189X020003002. 108

Salustri, F.A. and Rogers, D. (2008). Some thoughts on terminology and discipline in design. In *Undisciplined! Proceedings of the Design Research Society Conference 2008*. Sheffield, UK. 17

Satyanarayan, A. and Heer, J. (2014). Lyra: An interactive visualization design environment. *Computer Graphics Forum*, 33(3), 351–360. DOI: 10.1111/cgf.12391. 15

Scharein, R.G. (2015). KnotPlot [Computer software]. Vancouver, Canada: Hypnagogic Software. Available: http://knotplot.com/download/. 73

Schmid, U., Ragni, M., Gonzalez, C., and Funke, J. (2011). The challenge of complexity for cognitive systems. *Cognitive Systems Research*, 12(3), 211–218. DOI: 10.1016/j.cogsys.2010.12.007. 110

Schmidt, D. C. (1995). Using design patterns to develop reusable object-oriented communication software. *Communications of the ACM*, 38(10), 65–74. DOI: 10.1145/226239.226255. 19, 20

Schön, D. A. (1983). *The Reflective Practitioner*. New York: Basic Books. 20, 21

Schraagen, J. M., Chipman, S. F., and Shalin, V. L. (2000). *Cognitive Task Analysis*. Psychology Press. 122

Schulz, H. J., Nocke, T., Heitzler, M., and Schumann, H. (2013). A design space of visualization tasks. *IEEE Transactions on Visualization and Computer Graphics*, 19(12), 2366–2375. DOI: 10.1109/TVCG.2013.120. 110

Sedig, K. and Haworth, R. (2014). Interaction design and cognitive gameplay: role of activation time. *Proceedings of the first ACM SIGCHI Annual Symposium on Computer-human Interaction in Play*. ACM. (pp. 247–256). DOI: 10.1145/2658537.2658691. 116

Sedig, K., Klawe, M., and Westrom, M. (2001). Role of interface manipulation style and scaffolding on cognition and concept learning in learnware. *ACM Transactions on Computer-Human Interaction (TOCHI)*, 8(1), 34–59. DOI: 10.1145/371127.371159. 4, 108, 116

Sedig, K. and Liang, H. N. (2006). Interactivity of visual mathematical representations: Factors affecting learning and cognitive processes. *Journal of Interactive Learning Research*, 17(2), 179. 116

Sedig, K. and Liang, H (2008). Learner-information interaction: A macro-level framework characterizing visual cognitive tools. *Journal of Interactive Learning Research*, 19(1), 147–173. 118

Sedig, K. and Morey, J. (2002). A Descriptive Framework for Designing Interaction for Visual Abstractions. *2nd International Conference on Visual Representations and Interpretations, VRI'2002*, September 2000, Liverpool, UK. 4, 118

Sedig, K., Morey, J., Mercer, R., and Wilson, W. W. (2005a). Visualising, interacting and experimenting with lattices using a diagrammatic representation. *Studies in Multidisciplinarity*, 2, 255–268. DOI: 10.1016/S1571-0831(04)80046-7. 4

Sedig, K. and Parsons, P. (2013). Interaction design for complex cognitive activities with visual representations: A pattern-based approach. *AIS Transactions on Human-Computer Interaction*, 5(2), 84–133. 4, 19, 32, 48, 107, 108, 109, 111, 114, 118, 128

Sedig, K., Parsons, P., and Babanski, A. (2012). Toward a characterization of interactivity in visual analytics. *Journal of Multimedia Processing and Technologies – Special issue on Theory and Application of Visual Analytics*, 3(1), 12–28. 118, 128

Sedig, K., Parsons, P., Dittmer, M., and Haworth, R. (2013). Human-centered interactivity of visualization tools: Micro-and macro-level considerations. In *Handbook of Human Centric Visualization* (pp. 717–743). New York:Springer. 116, 118, 128

Sedig, K., Rowhani, S., and Liang, H. N. (2005b). Designing interfaces that support formation of cognitive maps of transitional processes: An empirical study. *Interacting with Computers*, 17(4), 419–452. DOI: 10.1016/j.intcom.2005.02.002. 116

Sedig, K., Rowhani, S., Morey, J., and Liang, H. N. (2003). Application of information visualization techniques to the design of a mathematical mindtool: A usability study. *Information Visualization*, 2(3), 142–159. DOI: 10.1057/palgrave.ivs.9500047. 4, 71, 72

Sedig, K. and Sumner, M. (2006). Characterizing interaction with visual mathematical representations. *International Journal of Computers for Mathematical Learning*, 11(1), 1–55. DOI: 10.1007/s10758-006-0001-z. 4, 110, 118

Sedlmair, M., Meyer, M., and Munzner, T. (2012). Design study methodology: Reflections from the trenches and the stacks. *IEEE Transactions on Visualization and Computer Graphics*, 18(12), 2431–2440. DOI:10.1109/TVCG.2012.213. 111, 119

Shannon, P., Markiel, A., Ozier, O., Baliga, N. S., Wang, J. T., Ramage, D., ... and Ideker, T. (2003). Cytoscape: a software environment for integrated models of biomolecular interaction networks. *Genome Research*, 13(11), 2498–2504. DOI: 10.1101/gr.1239303. 48

Shneiderman, B. (1982). The future of interactive systems and the emergence of direct manipulation. *Behaviour and Information Technology*, 1(3), 237–256. DOI: 10.1080/01449298208914450. 116

Shneiderman, B. (1996). The eyes have it: A task by data type taxonomy for information visualizations. *IEEE Symposium on Visual Languages*, pp. 336–343. DOI: 10.1109/vl.1996.545307. 110, 121

Shneiderman, B., Plaisant, C., Cohen, M., and Jacobs, S. (2009). *Designing the User Interface: Strategies for Effective Human-Computer Interaction* (5th ed.). Pearson. 119

Skyttner, L. (2005). *General Systems Theory: Problems, Perspectives, Practice* (2nd ed.). Hackensack, NJ: World Scientific. 24, 25, 121

Smiciklas, M. (2012). *The Power of Infographics: Using Pictures to Communicate and Connect with your Audiences*. Que Publishing. 124

Snodgrass, A. and Coyne, R. (1992). Models, metaphors and the hermeneutics of designing. *Design Issues*, 9(1), 56–74. DOI: 10.2307/1511599. 26

Spence, R. (2007). *Information Visualization: Design for Interaction* (2nd ed.). New York: Addison-Wesley. 36, 48, 124

Snyder, J. (2014). Visual representation of information as communicative practice. *Journal of the Association for Information Science and Technology*, 65(11), 2233–2247. DOI: 10.1002/asi.23103. 28

Stenning, K. and Oberlander, J. (1995). A cognitive theory of graphical and linguistic reasoning: logic and implementation. *Cognitive Science*, 19(1), 97–140. DOI:10.1207/s15516709cog1901_3. 4

Stolte, C., Tang, D., and Hanrahan, P. (2002). Polaris: A system for query, analysis, and visualization of multidimensional relational databases. *IEEE Transactions on Visualization and Computer Graphics*, 8(1), 52–65. DOI: 10.1109/2945.981851. 15, 48

Stolterman, E. (2008). The nature of design practice and implications for interaction design research. *International Journal of Design*, 2(1), 55–65. 15, 20

Svendsen, G. B. (1991). The influence of interface style on problem solving. *International Journal of Man-Machine Studies*, 35(3), 379–397. DOI: 10.1016/S0020-7373(05)80134-8. 116

Telea, A. C. (2015). *Data Visualization: Principles and Practice* (2nd ed.). CRC Press. 124

Thibodeau, P.H. and Boroditsky, L. (2011). Metaphors We Think With: The Role of Metaphor in Reasoning. *PLoS ONE*, 6(2), e16782. DOI:10.1371/journal.pone.0016782. 26

Thomas, J. C., Diament, J., Martino, J., and Bellamy, R. K. (2012). Using the "Physics" of notations to analyze a visual representation of business decision modeling. *Visual Languages and Human-Centric Computing (VL/HCC)*, 2012 IEEE Symposium on. IEEE. (pp. 41–44). DOI: 10.1109/VLHCC.2012.6344478. 28

Thomas, J. J. and Cook, K. A. (2005). Illuminating the path: The research and development agenda for visual analytics (No. PNNL-SA-45230). Pacific Northwest National Laboratory (PNNL), Richland, WA. 3, 12, 26

Tidwell, J. (2005). *Designing Interfaces*. Sebastopol, CA:O'Reilly. 19, 119

Tominski, C. (2015). *Interaction for Visualization. Synthesis Lectures on Visualization*, 3(1), 1–107. Morgan and Claypool. DOI: 10.2200/S00651ED1V01Y201506VIS003. 4, 110, 118, 121, 128

Tory, M. and Möller, T. (2004). Rethinking visualization: A high-level taxonomy. *Information Visualization, 2004. INFOVIS 2004. IEEE Symposium on* (pp. 151–158). IEEE. DOI: 10.1109/INFVIS.2004.59. 2, 3, 121

Trudel, C. I. and Payne, S. J. (1995). Reflection and goal management in exploratory learning. *International Journal of Human-Computer Studies*, 42(3), 307–339. DOI: 10.1006/ ijhc.1995.1015. 116

Tukey, J. W. (1977). *Exploratory Data Analysis*. Addison-Wesley. 10, 11

Tufte, E. R. (1983). *The Visual Display of Quantitative Information*. Cheshire, CT: Graphics Press. 2, 10, 11, 12, 13, 36, 48, 124

Tufte, E. R. (1990). *Envisioning Information*. Cheshire, CT: Graphics Press. 2, 10, 11, 13, 36, 124

Tufte, E. R. (1997). *Visual Explanations: Images and Quantities, Evidence and Narrative*. Cheshire, CT: Graphics Press. 10, 11

Tufte, E. R. (2006). *Beautiful Evidence*. Cheshire, CT: Graphics Press. 10, 11

Tversky, A. and Kahneman, D. (1973). Availability: A heuristic for judging frequency and probability. *Cognitive Psychology*, 5(2), 207–232. DOI: 10.1016/0010-0285(73)90033-9. 17

Ullman, D. G., Dietterich, T. G., and Stauffer, L. A. (1988). A model of the mechanical design process based on empirical data. *Artificial Intelligence for Engineering, Design, Analysis and Manufacturing*, 2(01), 33–52. DOI: 10.1017/S0890060400000536. 21

Vega, 2013. https://vega.github.io/vega/. 14

Visser, W. (1990). More or less following a plan during design: opportunistic deviations in specification. *International Journal of Man-Machine Studies*, 33(3), 247–278. DOI: 10.1016/ S0020-7373(05)80119-1. 21

Ware, C. (2008). *Visual Thinking for Design*. Morgan Kauffman. 11, 12, 127, 137

Ware, C. (2012). *Information Visualization: Perception for Design* (3rd Ed.). Waltham, MA:Morgan Kauffman. 2, 3, 11, 12, 27, 119, 127

Ward, M. O., Grinstein, G., and Keim, D. A. (2015). *Interactive Data Visualization: Foundations, Techniques, and Applications* (2nd Ed.). CRC Press. 36, 121, 124

Weinschenk, S. (2011). *100 Things Every Designer Needs to Know about People*. Pearson Education. 119

Wickham, H. (2009). *ggplot2: Elegant Graphics for Data Analysis*. Springer Science & Business Media. 14, 15

Wilensky, U. 1999. NetLogo. http://ccl.northwestern.edu/netlogo/. Center for Connected Learning and Computer-Based Modeling, Northwestern University. Evanston, IL. 48

Wilkinson L. (1999). *The Grammar of Graphics*. New York:Springer-Verlag. DOI: 10.1007/978-1- 4757-3100-2. 2, 3, 11, 12, 13, 121

Winn, T. and Calder, P. (2002). Is this a pattern? *IEEE Software*, 19(1), 59–66. DOI: 10.1109/52.976942. 19

Yau, N. (2012). *Visualize This!* John Wiley and Sons. 124

Yi, J. S., ah Kang, Y., Stasko, J. T., and Jacko, J. A. (2007). Toward a deeper understanding of the role of interaction in information visualization. *Visualization and Computer Graphics, IEEE Transactions on*, 13(6), 1224–1231. 4, 26, 118, 128

Zhang, J. and Norman, D. A. (1994). Representations in distributed cognitive tasks. *Cognitive Science*, 18(1), 87–122. DOI: 10.1207/s15516709cog1801_3. 2, 9, 108

Zhang, J. and Patel, V. L. (2006). Distributed cognition, representation, and affordance. *Pragmatics and Cognition*, 14(2), 333–341. DOI: 10.1075/pc.14.2.12zha. 108

Zhou, M. X. and Feiner, S. K. (1998). Visual task characterization for automated visual discourse synthesis. *Proceedings of the SIGCHI Conference on Human Factors in Computing Systems* (pp. 392–399). DOI: 10.1145/274644.274698. 110, 121

Zins, C. (2007). Conceptual approaches for defining data, information, and knowledge. *Journal of the American Society for Information Science and Technology*, 58(4), 479–493. DOI: 10.1002/asi.20508. 121

About the Authors

Kamran Sedig is an Associate Professor in the Department of Computer Science and the Faculty of Information and Media Studies at Western University, Canada. Before joining Western, he was a Research Staff Member at the IBM T.J. Watson Research Center in New York. He received his Ph.D. in Computer Science (Interactive Visual Games for Learning Math) from The University of British Columbia, where his research was nominated for the Governor General's Gold Medal for best Ph.D. dissertation. His M.Sc. was in Computer Science (Artificial Intelligence) from McGill University. His research interests are in human-information interaction, joint human-computer cognitive systems, visual analytics, interactive reasoning with visualizations, visual interface design, and interaction and task design for complex cognitive activities. Sedig's research uses a human-centered approach with a particular focus on human cognitive needs and activities. He is the director of the Insight Lab (insight.uwo.ca).

Paul Parsons is an Assistant Professor in the Department of Computer Graphics Technology at Purdue University. In 2013 he received his Ph.D. in Computer Science from Western University, Canada, with a focus on the design of interactive visualizations and interfaces that support cognitive activities. His research interests are in the areas of information visualization and visual interfaces, human-computer and human-information interaction, cognitive and learning technologies, and human-centered computing and design. He is particularly interested in understanding and supporting cognitive processes and activities of both users and designers.